TABLE OF CONTENTS

FOREWORD

Because of the increasing role of natural gas in many countries and its importance as an alternative to oil use, there has been a growing international interest in natural gas issues. Gas has become a more internationally traded commodity and there has been an increasing reliance in Western Europe and in Japan on imports of gas. The International Energy Agency (IEA) published in 1982 a study entitled "Natural Gas Prospects to 2000" and concern about security of supply led the IEA's Governing Board meeting at Ministerial level on 8th May 1983 to adopt a series of conclusions on natural gas (Annex I). This was followed up when the Governing Board met at Ministerial level on 9th July 1985 (Annex II).

The present study brings the 1982 study up to date and analyses issues and policies relevant for the development of natural gas demand and supply in the IEA countries in the remainder of this century and into the first decade of the next. The study has been prepared by the IEA's Secretariat in co-operation with gas and energy experts from the Member countries. It is published on the responsibility of the Executive Director of the IEA and the results and views found in this study do not necessarily reflect those of IEA Member Governments.

The present study addresses policy questions in the Member countries of the IEA. However, because supply of and demand for gas in the member countries of the IEA in Europe cannot be considered in isolation from the West European gas situation as a whole, the projections presented for gas supply and demand cover the whole OECD area and not only the Member countries of the IEA.

Note on Conversion Factors

In this study, one cubic metre of natural gas is defined to equal 9 500 kilocalories (39.7746 megajoules) on a gross calorific value (gcv) basis.

One million metric tons of oil equivalent (Mtoe) is defined as 10^{13} kilocalories, and is expressed on a net calorific value (ncv) basis, consistent with OECD statistical practice.

Accordingly, to convert from one billion cubic metres (gcv) of gas (10^9 cubic metres), to million metric tons of oil equivalent (ncv), divide by 1.148.

Detailed conversion factors are given in Appendix IV.

Terminology and Constituents of Natural Gas

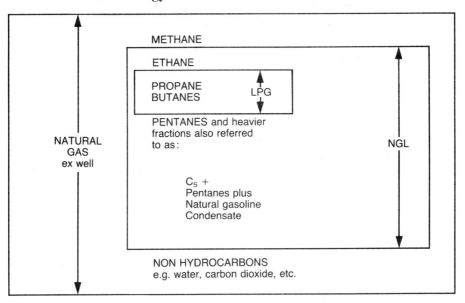

LNG = liquefied natural gas   NGL = natural gas liquids
LPG = liquefied petroleum gas   SNG = synthetic (or substitute) natural gas
Source: Shell International Gas Limited.

I. INTRODUCTION

Natural gas has a long history as an energy source. It was, however, not developed on any important scale before this century and in the early start of the industry, consumption was developed close to the producing fields. Although several countries started production of natural gas after World War I, the United States' market remained by far the most dominant in the world. In 1945, for example, total world production was about 120 billion cubic metres (bcm) of which the United States accounted for 92%. Development of the gas markets in other parts of the world took place in the 1960s after significant new discoveries of gas reserves. In some countries, the penetration of natural gas to small-scale consumers was facilitated in a first phase by the already existing town gas networks which originally had been constructed for distributing coal-derived gas to consumers. In other countries completely new grids have been laid from the very beginning of the natural gas industry. The expansion of gas use has, particularly since the 1960s, been followed by increasing international trade and long distance transport of gas in order to supplement production from local reserves. Gas is for example piped from the Norwegian North Sea to continental Europe and to the United Kingdom, liquefied natural gas (LNG) is shipped to Japan over thousands of kilometres, and Soviet gas is piped from Siberia to the western parts of the Soviet Union, to Eastern European countries and to Western Europe.

Natural gas use has been and still is one of the important alternatives to oil. It is clean compared to other fossil fuels and has in many applications an inherent advantage in use because of its capability of being easily regulated in burning with high efficiency. On the other hand, transportation and distribution of natural gas require investments in transmission or in liquefaction facilities. The tendency to obtain a larger share of supplies from more distant sources of supply implies higher transport costs as a part of the total costs of bringing natural gas to the market. Furthermore, the real costs of finding and developing natural gas reserves have been increasing in many parts of the world, because of the relatively high costs of offshore developments and the increasing costs of drilling deeper into the ground.

In spite of these developments, total natural gas use in the OECD area has increased from 607 Mtoe in 1970 to 702 Mtoe in 1984. The share of gas in total primary energy requirements has remained at about 19% in the same period. The interest in natural gas as an alternative to oil has been maintained and the oil price increases in 1973-74 and again in 1979-80 made it economically attractive to bring on stream new sources of supply with high development costs and to expand the markets in many countries of the OECD. Several completely new markets have been opened and trade in gas between the OECD countries themselves and with non-OECD countries has grown in importance. Gas is, however, because of the costs of transportation, still much less traded than oil or oil products and although regional gas markets have become more integrated, there are no direct links between regions. There are strong indirect links between markets simply because gas in most parts of the world is in competition with oil products, for which the international markets in turn have become increasingly integrated over the last decade. Differences in gas market developments between regions of the OECD are, however, still significant.

In the European countries of the OECD, the natural gas market was built up gradually from a very small size in the early 1960s and the market continued to grow, although at declining rates until 1979. The economic recession after 1979 and also some erosion of the competitiveness of gas against oil products depressed gas demand in subsequent years but demand started again to grow in 1983 and the growth continued into 1985. The share of gas in total primary energy demand has increased since the early 1970s from about 6% to 15% (see Figure I-1). There has been some shift between sectors. The residential/commercial sectors[1] have increased their share in total gas demand and the shares of the industrial and electricity generation sectors have decreased. Although there has been some revival of industrial gas demand in recent years, the trend towards a higher share of residential/commercial demand has continued.

---

1.   Includes agricultural and public sector use throughout this report.

Figure I-1
Total Primary Energy and Natural Gas Demand
OECD Europe

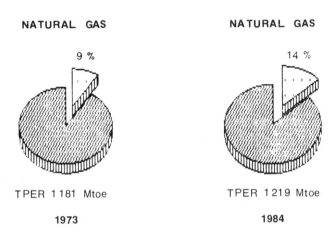

NATURAL GAS	NATURAL GAS
9 %	14 %
TPER 1 181 Mtoe	TPER 1 219 Mtoe
1973	**1984**

Source: IEA/OECD Energy Statistics.

The same sectoral trend is apparent in North America, but there is, apart from this common feature, a remarkable divergence between the gas market development in the large United States' market and in Western Europe. In the United States, gas demand reached an all time peak in 1972 and declined by about 25% by 1983. There has at times been a lack of supplies sufficiently competitive with oil in order to maintain the position of gas. The interaction between demand and supply has been distorted by regulation of prices at the well head, and parallel to the loss of market, a surplus supply potential has built up. The overall economic development played a role not only in depressing total energy demand but also in reducing gas demand. In Canada, which has a much younger gas market than the United States, the development in total gas demand has followed lines close to those described above for OECD Europe. In the most recent years, regulatory developments in the United States and also the existence of surplus production capability (the "gas bubble") have led to a more competitive climate which, concurrent with an economic upswing, led to an increase in gas demand in 1984. Gas use, nevertheless, remains considerably lower than it was a decade earlier (see Figure I-2).

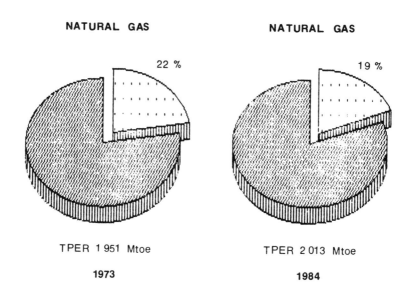

Figure I-2
**Total Primary Energy and Natural Gas Demand
North America**

NATURAL GAS

22 %

TPER 1 951 Mtoe

1973

NATURAL GAS

19 %

TPER 2 013 Mtoe

1984

Source: IEA/OECD Energy Statistics.

The Pacific markets for gas have been by far the most dynamic of the markets in the OECD area in the last decade (see Figure I-3). Indigenous reserves have been developed in Australia and in New Zealand. In Japan, gas use has increased sharply as a result of the coming on stream of a number of liquefied natural gas (LNG) projects for supplying primarily the Japanese electric utilities. This has been part of a deliberate Japanese policy to reduce the country's dependence on imported oil; and the increased use of gas has also made a major contribution to the reduction in air pollution from electricity generation. The planned process of substitution in electricity generation and the nature of LNG trade, dominated as it is by long-term planning and contractual commitments, implies that the development in gas use in the Pacific region of the OECD has been much less influenced by short-term fluctuations than has been the case in other countries or regions of the OECD.

In several of the major and "mature" gas markets, consumption of gas declined in the first years after the oil price increases in 1979-80. The main reason for the decline in gas demand was the economic recession which followed in the wake of the oil price increases and the consequent contraction in total demand because of lower industrial activity and because of increased conservation efforts for all kinds of energy including natural gas. In some markets, gas also lost ground because prices of gas at least temporarily were out of line with competing oil product prices. Many industrial users of gas switched from gas to oil use when heavy fuel oil became priced more attractively than gas. The most recent years, however, saw a clear reversal of this development.

Figure 1-3
Total Primary Energy and Natural Gas Demand
OECD Pacific

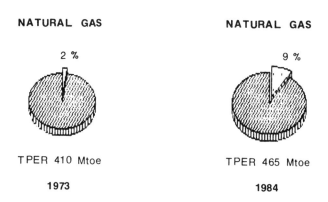

NATURAL GAS	NATURAL GAS
2 %	9 %
TPER 410 Mtoe	TPER 465 Mtoe
1973	1984

Source: IEA/OECD Energy Statistics.

More competitive pricing of gas in combination with higher economic activity led to substantial increases in gas consumption in 1983 and 1984 and the growth in overall gas consumption continued, although at a lower rate, in 1985.

Despite the very wide variety among gas markets, there is one common feature in recent developments in many gas markets: because growth of total energy and gas demand has been lower than was earlier expected, contracted and developed supply potentials have tended to exceed demand at ruling prices. This development towards a "surplus" situation in many markets has taken place parallel to similar developments in other energy markets: coal, electricity and not the least, oil. The result has been more competition between gas and other fuels and in some markets between gas suppliers. This is a situation which could continue for some time and in a sense the natural gas industry is at a crossroads. The last decade has seen the completion of a series of major new supply projects implying heavy investments in supplying countries inside and outside the OECD and in consuming countries. Continued development of new supplies is necessary to supplement declining production from existing fields and in order to meet any increase in demand. Because of the long lead times involved, especially in offshore developments, the investment climate today and in coming years will have a decisive influence on the gas market in the 1990s.

The supply situation differs among the regions. In Europe the supply picture is dominated by long-term contracts, some of which are running into the next century. The long-term supply prospects have improved. In recent years there have been upward revisions in the estimates of the feasible production available for local use in continental Europe and in the United Kingdom. The contracts for exports of Dutch gas have been extended and now guarantee supplies into the first decade of the next century. Negotiations for exports of new Norwegian supplies, which could be available in the second half of the 1990s, are well under way. In North America, supplies are to a much greater extent governed by short- and medium-term arrangements. There has developed a situation of oversupply in the United States market, a situation which could last for several years, but in the longer term supplies could decrease. Although further increases in exports from Canada to the United States are feasible, there is a risk, in particular in the current uncertain situation on the oil markets, that supply constraints in the longer term will limit the size of the United States gas market which could lead to an increased use of oil and add to pressures on world oil markets in the 1990s. In the Pacific, the important Japanese markets look reasonably assured well into the 1990s. Supplies of Australian liquefied gas for the Japanese market have been contracted and the now lower than earlier projected gas demand in the 1990s has postponed the need for major additional supplies.

According to projections undertaken by the IEA Secretariat, total gas demand in the OECD countries could increase substantially in the 1990s and after the turn of the century. The previous projection of natural gas demand presented by the IEA's Secretariat in the study "Natural Gas Prospects to 2000" published in 1982 saw total OECD demand growing to 962-989 bcm in 1990 and 954-1 089 bcm in 2000. The present study has a less buoyant view of growth potential. The instability of today's energy market makes projections particularly uncertain but gas demand in OECD countries seems unlikely to be much above 1 000 bcm in the year 2000. After the turn of the century, gas demand could continue to grow but at a rate which is lower than that thought feasible for the remainder of the 1980s and the 1990s. These projections are optimistic in the sense that they assume that gas supplies will be available at prices competitive with other fuels and in particular with oil products in order to meet the projected increase in demand. In the short- to medium-term perspective, this seems a reasonably safe assumption, granted that supplies over the coming five to six years in the main will come from fields already in production or where the development and investments are already well under way. The supply in the short- to medium-term should therefore not to any significant degree be influenced by oil price developments unless regulation impedes adaptation of gas prices to developments in oil markets. For the longer term, however, the supply prospects tend to be much more closely related to oil price developments. A prolonged period of relatively low oil prices carries the risk of reducing the exploration and development activity which is necessary in order to increase and in some areas even to maintain supplies of gas. This could have the result of increasing the reliance on oil use from the late 1980s and in the 1990s.

II. SUMMARY AND CONCLUSIONS

A. Summary

(i) Organisation of the Natural Gas Industry (Chapter III)

In most OECD countries, governments have taken an active role in influencing the gas industry. In the United States, for example, regulatory agencies have been an established part of the industry's operations. In several Western European countries state-owned corporations are involved in the industry. In Japan, the Government has applied tax benefits and subsidised loans for the development of LNG trade.

In Europe, only a relatively few companies are involved in production and transmission; purchasing and transmission of gas is sometimes in the hands of a dominant buyer. The important trade among the European countries and with outside suppliers is governed by long-term contracts. In the United States, production is undertaken by several thousand independent producers, by major oil companies and by some of the major gas pipeline companies. There has been a transition towards greater competition among producers because of changes in the regulatory environment and the existence of surplus supplies. There is an increasing reliance on short-term contracts. In Japan, the gas supplies are provided by the linked operation of major consumers, which are electric utilities, city gas companies and one steel company. The supplies are governed by long-term contracts.

(ii) Demand Prospects to 2010 (Chapter IV)

Possible developments in gas demand have been evaluated by the IEA Secretariat on the basis of two assumed paths for economic growth and oil prices. In both cases oil prices are assumed to decline from average 1985 levels, but then to increase in the 1990s and after the turn of the century. On this basis and combined with assumptions about the growth in the economy as well as several other factors, demand for gas is projected to grow slightly less than total energy demand, implying a long-term growth rate in total gas demand of between 0.7% and 1.3% per annum from 1984 to 2010. The situation varies among countries and regions. In the European OECD countries as in the United States, the residential/commercial market is by far the most important in total gas demand and its share is in both areas projected to increase. In Canada, the industrial and residential/commercial markets have about equal weight and are projected to develop very much in line with each other. In Japan, electricity generation is the most important consuming sector and is projected to remain so.

(iii) Reserves and Supply Prospects to 2010 (Chapter V)

The analysis of demand takes no account of possible supply constraints: it assumes that supply is available at prices competitive with other fuels. In a geological sense, reserves of gas are plentiful, but reserves of gas do not constitute a supply of gas. Future supply of gas will be a function of gas committed to the market by suppliers from reserves which can be developed commercially, that is, so-called active reserves. Any estimate of active reserves or any other category is fraught with difficulty. It is expected that based on existing known reserves, there is no shortage of natural gas in the foreseeable future. In mature gas provinces of the world, however, maintaining an established level of supply will tend to require an increasing level of investment effort and cost.

The costs of delivering gas include, apart from the production costs, the costs of transport and the costs of providing supply in a manner acceptable to the buyer. In the United States, there is a tendency towards higher costs of production because of deeper drilling, and transport costs could increase in the longer term, because it may be necessary to transport gas over longer distances. Also in Europe present and future trunk transmission investments, as well as production costs, are critical elements in the costs of supplying from various sources. In

Japan, the cost of gas supply is heavily conditioned by the high fixed costs of the various LNG chains, which have been constructed to serve the market.

In the European OECD countries taken as a group, supply of gas for use in the country of production could be sustained at about the present level of 110 bcm. Contracted exports from Norway will decline in the 1990s, but new contracts and developments in Norway could offset the decline. Other supplies additional to those already contracted include possible extensions of contracts with the Soviet Union and Algeria after 2000 or additional volumes from these countries before that. Other sources of gas could come from Nigeria and Qatar where LNG projects are still under review. In the 25 year timescale of this study, there is of course a significant possibility that new sources of gas, within or outside Europe, could be discovered and developed for the European market. For instance, the present levels of supplies of gas from Norway and the Soviet Union were not foreseen nor readily foreseeable 25, or even 20 years before they started flowing to European markets. In the United States, the major possible sources, in addition to conventional and tight formation production in the lower 48 States, are continued imports from Canada, imports from Mexico, imports of LNG from diverse sources, transport of gas from Alaska to the lower 48 States and non-conventional sources of supply like synthetic natural gas. After 1990, conventional gas production in the lower 48 States is likely to be in long-term decline. The long run decline in conventional production in the lower 48 States could perhaps be partly offset by an increase in exports from Canada, but there is some uncertainty about the long run production capacity also in Canada. For Japan, it is probable that volumes under existing contracts plus small quantities of indigenous production would be sufficient to cover demand in 1990 and could be sufficient through to 2000 by renewing or extending existing contracts. New supplies are contracted from Australia and other sources of supply have been discussed.

The critical issues are what level of price gas companies can pay and still be able to sell the gas in a competitive market and if these prices result in an adequate return to encourage exploration for and production and supply of gas. A simplified calculation for supplies to OECD Member countries in Europe indicates that in the oil price scenarios assumed for the study, major new North Sea field developments are economically feasible; LNG transported from the Middle East or from West Africa also appears to be a possibility in the long run.

(iv) Environmental Effects of Natural Gas Production, Transport and Use (Chapter VI)

Compared to other fossil fuels, natural gas has distinct advantages as a clean-burning fuel. Gas is normally cleaned at the point of production for impurities like sulphur components. Nitrogen emissions can be kept at a relatively low level by appropriate burning controls and techniques, for gas as for other fossil fuels.

(v) Storage and Contingency Planning (Chapter VII)

Storage of gas has an important function of matching seasonal variations in demand in combination with available flexibility in supplies and in demand (for example, interruptible customers). Storage can also play an important role in a disruption of supply. Present plans for 1990 indicate that storage capacities will be expanded in OECD Europe in response to a deterioration in load factors because of an increase in the share of residential/commercial demand in total demand. An increase in storage capacities is also planned in Japan. In the longer run, in Europe there is likely to be a requirement for either a considerable flexibility in the deliverability profile of the next major tranche of gas or further expansion of storage systems in order to maintain the same degree of security of supply. Other arrangements could also be needed. These could include improved regional co-operation and more physical flexibility in the transmission system in Europe.

B. Conclusions

In purely physical terms, the resource base suggests that no scarcity of natural gas supplies could develop. The existence of large gas reserves does, however, only result in supplies if the commercial possibilities exist for developing the reserves and bringing them to the market. Gas use has environmental advantages compared to other fossil fuels and is an important alternative to the use of oil in the OECD countries. There is, therefore, a policy interest in securing that new OECD supplies are developed in a timely manner in order to match the development in demand.

The situation differs among regions: the consuming centres in Western Europe are surrounded by several alternative sources of supply, including the North Sea, Soviet Union and Algeria, and there will be a need to maintain imports from non-OECD sources of supply in the long run. In order to maintain a balanced pattern of supply in the late 1990s, it is important that new Norwegian supplies now under consideration are developed. In a situation of declining energy prices, and of short- to medium-term over supply, there is a risk of postponement of investment decisions because of increased uncertainty. Some progress has been made along the lines laid out in the IEA Ministerial Conclusions of May 1983 and July 1985, but it remains essential that the security aspect of the supply pattern as well as other measures to improve security such as storage and regional co-operation, be given continued attention by Member countries of the IEA, and by appropriate bodies and international organisations. The European Community, among others, is paying attention to this.

In the important United States' market, regulatory developments have contributed to create a surplus supply of gas after having contributed to shortages in the mid-1970s. The market is now becoming increasingly competitive. Supply restraints and increasing costs of developing and transporting gas to the market are in the longer term a matter of concern. Much will depend on the development in exports from Canada which again depend on the balance between demand and production in that country. Other sources of supply are feasible but there is a risk that supply constraints in the 1990s could lead to a shift from gas to oil use particularly in the industrial energy market. Further deregulation of the United States gas industry would be likely to increase its responsiveness to changing market circumstances and should be pursued.

In Japan, existing contracts for imports of LNG appear likely to be adequate to cover demand through the mid 1990s. Extensions of existing contracts are possible and new projects are also still under consideration. In volume terms, the Japanese market appears to be reasonably secured but some problems of a commercial nature need to be resolved as the use of gas for electricity generation moves from baseload towards medium load.

In order to preserve the role of gas in energy markets, it is of importance to price it competitively with other fuels. The pricing practices of some non-OECD exporters have been at variance with this basic principle and pricing regulations have also in some of the consuming countries created distortions affecting both demand and production. On the supply side, prices paid to producers must also be adequate to promote exploration for and development of new gas resources if gas is to fulfill its potential in the energy markets of the OECD. It is of importance to establish sound pricing principles both in international trade and nationally and it is an area where most governments have an important role to play.

There are many uncertainties facing the gas industries in the OECD countries. Future developments in demand are determined by a number of factors, which are impossible to forecast with any degree of precision, particularly in the current situation on the oil markets. The gas industry has been able in the past to adapt to changing circumstances with a high degree of flexibility, which has been in part due to the extent of international integration and interdependence in the industry. It is important that OECD Member governments maintain policies which enable the gas industry to continue to have this flexibility, and that gas market developments are kept under review at both the national and international level.

III. ORGANISATION OF THE NATURAL GAS INDUSTRY

A. Origin and Location of the Industry

The major gas industries of the IEA countries have grown up, as elsewhere in the world, where reserves of natural gas have been proved and readily exploited. Australia, Canada, the Netherlands, the United Kingdom and the United States have developed industries based essentially on domestic resources. In the United States and the United Kingdom, consumers have benefitted from supplementary supplies from neighbouring countries (Canada, Mexico and Norway, respectively). In the case of the United States, imported gas could compete because United States' regulatory controls held some domestic production at prices below the market-clearing price, which permitted higher priced imports to be rolled in to supply portfolios. In the United Kingdom, alternative indigenous supplies could not be made available in the time scale required to meet projected demand and there was an option to develop on a unitised basis a large gas field which lay across the median line between the British and Norwegian sectors of the Continental Shelf.

New industries are now developing in Denmark, Ireland, and New Zealand based on domestic reserves discovered in the 1960s and 1970s. Austria, the Federal Republic of Germany, Italy and Japan, which for many years have had small, locally-based domestic natural gas indus-tries, have taken advantage of opportunities in international trade to build natural gas into an important element in national fuel supply. The natural gas industry in Belgium and Luxembourg has developed on the basis of the proximity of the Netherlands' reserves. Natural gas makes a much smaller contribution to fuel supply in Spain and Switzerland, where imports must be brought in over longer distances. Recent discoveries of indigenous reserves have allowed production to begin on a small but locally significant scale in Spain.

In the European countries of the IEA, active reserves can support a substantial rate of domestic supply well into the next century. Government policies in Europe have a major impact on the rate at which domestic reserves are developed, through licencing procedures and depletion objectives, as well as through the various production tax systems in force. The balance between obtaining a desirable rate of supply from domestic European and outside sources, will present a continuing challenge to policy. In the Pacific region, a lack of substantial domestic reserves means that Japan, the largest gas-consuming country, will continue to rely on international trade for the supply of natural gas. In North America, the United States may face a supply constraint, unless as yet undiscovered and even unconventional resources can be converted into active reserves on a large scale. The United States could once again become a major player in international gas trade. This could require major changes in the economic and policy environment. Political choices are likely, therefore, to continue to play a role in the prospects for gas supply in all the main regions of the OECD.

B. The Role of Governments

In most IEA countries, governments have taken an active role in influencing the development of many aspects of the natural gas industry. In addition to obvious areas of public concern, such as the establishment and monitoring of safety requirements, or the licencing of production and transmission facilities and their location, governments have taken on a wide range of controlling functions relating to matters of a more general economic or strategic nature.

In the United States, regulatory agencies have been an established part of the industry's operations at both state and federal level for many decades. Transmission company rates of return have been subject to government control since the Natural Gas Act of 1938 and, following a decision by the Supreme Court in the Phillips' case in 1954, wellhead prices for gas sold between States were also subject to government approval. The price for gas sold within a State (the "intrastate" market) remained freely determined by negotiation between buyers and sellers until 1978. The Natural Gas Policy Act of 1978 established wellhead price ceilings for certain categories of gas in both intrastate and interstate markets according to the production characteristics of the gas, and stipulated that there should be gradual progress towards decontrol of these prices according to phase-out provisions for the price-ceilings in some of the categories. By the 1st January 1985, these phase-out provisions had taken their full effect, so that between 55% and 60% of United States gas was no longer subject to federal price legislation at the wellhead. Without further legislation, gas whose price is still regulated will remain so until it is exhausted. In recent years, lower oil prices, legislative developments, and certain key rulings by the Federal Energy Regulatory Commission, notably regarding minimum bills in purchase contracts and authorisations for pipelines to transport gas on behalf of third parties, have contributed to promoting a more competitive environment. A strong buyers' market has emerged in the United States and the combination of all the above factors has created a more market-driven and less government-influenced industry. The transition has been dramatic, and strong competitive forces are still at work restructuring the industry. Its ultimate effects remain uncertain, as the industry now operates in an environment which is at the same time part-competitive and part-regulated. The United States' Secretary of Energy has announced his intention to introduce legislation to deregulate the natural gas market in the 1986 Congress. Administrative changes have also been proposed to give higher returns on currently price-controlled old gas within the existing legislative and regulatory framework. Comprehensive deregulation would result in more flexible contracts between producers and purchasers of gas and in a natural gas market which is more responsive to changing conditions. It could tend also to promote reliance on United States lower cost energy and gas resources.

In most Western European countries production of natural gas is the concern of private companies, but state-owned corporations are also involved in this phase of operations, for example, ÖMV in Austria, Agip in Italy and the British Gas Corporation (BGC) in the United Kingdom. Government-owned Statoil participates in most Norwegian production licences. In many European countries, governments have taken a direct ownership role in the transmission of natural gas, through the medium of state corporations or nationalised holding companies. This has been the case in Austria, Denmark, Ireland, Italy, Spain, and the United Kingdom although in Great Britain, the BGC is to be sold to the private sector. In other countries, there is a mixture of state and private holdings in a company with sole purchase and distribution rights (as in the Netherlands), or with sole transportation rights (as in Belgium). Transmission of gas in Germany is divided up into regional concessions by private law agreements, but there is only one indirect federal government holding in a transmission company and the Government does not influence company policy. At the local distribution level, government involvement is normally limited to local authority interests in municipal distribution companies. This is the pattern in all IEA countries in Europe except for the United Kingdom, where distribution of gas to most of the country is the responsibility of the BGC.

Western European governments in all countries except the Federal Republic of Germany, Sweden and Switzerland take responsibility for establishing or approving the level and structure of tariffs to gas consumers. Such powers have been used in various ways in the past for objectives as diverse as counter-inflation policy, the promotion of desired market shares for different fuels in an overall energy plan, and for reducing costs to industries which may be vulnerable to overseas competition. The Federal Republic of Germany, Sweden and Switzerland are notable exceptions where there has been no government control of prices at any stage in the operation of the industry from production at the wellhead or import at the border to distribution to the consumer. In Germany, the Federal Cartel Office, however, checks the level of consumer prices to confirm that there is no abuse of market position.

In Western Europe, the United States and Canada, governments must give approval for the terms of proposed import and export contracts before international trade can take place. This function took on an increasing importance in Europe in the 1970s and 1980s as the number and size of such deals became larger. Foreign exchange and counter-trade implications in connection with imports of gas from the Soviet Union and Algeria have also increased as the unit value of natural gas rose in line with oil prices. Questions of the security of supply arose in a number of European countries and an informal understanding was reached between governments and utilities on the need for diversification of supplies.

As well as the national supervision of the gas industry in various Western European countries, there is an important element of international integration. This is provided not by governments nor by multinational institutions, but by the oil and gas industry itself. Formal layers of international integration include common ownership of certain transnational pipelines, and, in some cases, common shareholders. Multinational oil companies and private sector transmission companies in the Federal Republic of Germany play a major role here. This is discussed in section C (i) below.

In Japan, the Government's main area of involvement has been to give impetus through tax and subsidised loan programmes to the development of LNG trade with many different supplying countries as a means of helping to diversify the economy, and particularly the electricity sector, away from oil. The business itself is in the hands of the privately-owned electricity and gas utilities based in the major cities, with project development assisted by major trading houses. Tariffs are sanctioned by the Government on a cost-of-service basis, in line with the Electric Utility Industry Law and the Gas Utility Industry Law, and are therefore not very flexible vis-à-vis energy market fluctuation.

C. The Institutional Background

(i) IEA Europe

In European IEA countries, relatively few companies are involved in production and transmission of natural gas. Production is dominated by the major national (Agip, ÖMV, Statoil) and multinational oil companies (British Petroleum, Conoco, Esso, Mobil, Phillips and Shell). Some of these companies have important interests in transmission companies and transboundary pipelines, and were intimately involved in setting up natural gas marketing on a national scale in many European countries. International co-operation in transmitting gas is facilitated by joint ownership of major trunk pipelines — for example, SNAM of Italy and Ruhrgas of Germany jointly own the Trans Europa Naturgas Pipeline (TENP) which supplies Dutch gas to Italy and Switzerland, and Ruhrgas, Gaz de France, and ÖMV of Austria each own part of the Mittel Europäische Gasleitungsgesellschaft (MEGAL) which connects Czechoslovakia with France and Germany for the supply of Soviet gas. Producing companies, Shell and Esso, also have interests in trunk pipelines in the Federal Republic of Germany, which has a pivotal geographical position as the point of import for three of the four major continental European suppliers of gas (the Netherlands, Norway and the Soviet Union). This aspect of private sector involvement in the German gas industry provides one important area of opportunity of international integration.

Gas purchasing for transmission and reselling is sometimes in the hands of a dominant buyer — as in Italy, the Netherlands and the United Kingdom, — and where there are a number of buyers, as in Germany, long-term contracts may be in force which guarantee some pipeline companies access to part of any reserves discovered and developed by producers. For international purchases, continental European transmission companies sometimes form a consortium to conduct negotiations. Consortia may also redistribute gas supplies among their members if evolving demand conditions make this necessary. This is an important risk-sharing mechanism which contributes to security of supply for individual consortium members. Co-operation in consortium arrangements also provides an important, although informal, means of ensuring close contacts among transmission companies as well as between transmission companies and some of their suppliers. These contacts are used to reinforce the understanding of mutual shared interests in long-term reliability of supply and offtake. They are sometimes also used to ease shorter-term operational problems which may be faced by individual consortium members, but which the group as a whole can accommodate more easily. In general, private law contracts and informal relationships play a very important role, not only in the supply of gas but also in establishing the framework in which industry operations take place. Thus, the framework of the European gas industry remains generally less regulated by government authority than the framework in which the United States' industry must operate.

In the United Kingdom, new legislation is in progress to transfer the British Gas Corporation to the private sector. British Gas is at present close to being the sole purchaser and distributor of natural gas in the United Kingdom, although under a law introduced in 1982, its legal monopoly was ended. The privatised company will have freedom to act as a commercial business. United Kingdom policy on gas imports and exports will also be made more open to market forces. The privatised company will be able to import gas and there will no longer be a presumption against gas exports. Regulation will be limited to downstream supply and pricing of gas to tariff (mainly residential) customers.

(ii) United States

The United States' gas industry is divided into three distinct segments — corresponding to the production, transportation and distribution phases of gas supply operations. The operations of the production phase are undertaken by several thousand small independent producers, by major oil companies, and by the producing arms of some major gas pipeline companies. In the second phase, gas is transported and stored through intrastate and interstate pipelines by pipeline companies, a relatively small number of which account for the major part of gas transmitted and sold. In 1983, 13 pipeline companies transported 69% of all gas consumed in the United States, either having taken title to the gas for resale (53%) or on behalf of third parties (16%). The third main activity is distribution of gas, in which about 1 700 distribution companies are involved. Roughly a third of these are municipally or publicly owned, two-thirds are privately owned. Distribution companies normally operate within franchised service areas and under service obligations, and designated service areas do not overlap even where more than one company operates in a city area. Their operations are regulated at state or local level.

The production phase of operations is already a highly competitive area. The transition towards greater competition through which the United States' gas industry is now passing essentially relates to the transmission and distribution phase of operations, even though the most visible sign of change under the Natural Gas Policy Act of 1978 related to the decontrol of certain wellhead prices (that is, in the production phase). In the mid 1980s, pipeline companies are being driven by changes in regulations and by competitive pressures to offer transportation service for gas owned by others, rather than exclusively to take title to and resell gas. The new legislative and regulatory environment and the existence of surplus supplies have enabled some large-volume consumers, or distribution companies in some cases, to "shop around" for lower cost supplies; in other cases, they may be able to pressure their traditional suppliers into providing gas at lower cost. The pipeline companies in turn have pressured producers into reducing contract prices, or into changing the mix of low-cost/high-cost gas available under different contracts. In the extreme case, many pipeline companies have reneged on contracts or extracted reductions in take-or-pay commitments, which have on average been honoured in the range of $0.10-$0.15 in the dollar in 1984 and 1985[1]. One major result of these developments will be to refocus the

---

1. It is difficult to put a precise value on the degree to which contracts have been honoured, because many settlements have included non-money elements such as release of committed gas for sale to others, linking up of new reserves to pipeline systems, changed billing and payment systems, etc.

lead responsibility for gas supply adequacy on the local distribution companies and state regulators, and away from the pipelines and Federal Government, who have traditionally taken most of this responsibility.

Under these circumstances, pipelines and producers are increasingly relying on a mix of long-term, short-term, and spot contracts. In 1984, about 12% of gas was sold under spot contracts with either one year or less duration, or two year duration with at least quarterly and usually monthly price reopeners. In 1985, the proportion of spot gas had risen to about 17%. Perhaps even more significantly, long-term contracts are themselves changing in form. In some cases, contracts are really only binding for three years with optional two-way "market out" clauses thereafter. This is similar to European contracts which usually allow periodic renegotiation of price and/or volume terms for commercial problems, and which contain "hardship" or "unforeseen circumstances" clauses to cope with difficulties of a technical and commercial nature. Long-term contracts in the United States are also becoming shorter in term, and virtually all new contracts are being written with flexible pricing terms. To complement the operation of spot gas markets, the New York Mercantile Exchange (NYMEX) has a proposal for natural gas futures contracts pending before the Commodities Futures Trading Commission (CFTC). Gas deliverable in the Texas intrastate market would be delivered at two separate fields, and interested pipeline companies are planning a joint venture to ensure transportation. The prospects for this and other market and contractual developments will depend on the eventual framework established by pipeline company responses to federal regulatory initiatives on the rules governing taking title to and transportation of gas.

The implications of the trend towards the use of a mix of short, long, spot, and possibly futures contracts are difficult to assess. It has been welcomed as a means to prevent the wide swings between shortage of supply (as in the period of curtailments in the 1970s) and surplus deliverability with rising end-user prices (as in the emergence of the gas "bubble" since 1980). On the other hand, there are concerns that the supply outlook is poor and that the reluctance of pipelines to make purchase commitments will contribute to depressing exploration and development activity, as the structural changes imply a higher degree of risk and financial exposure at the production end. The changing institutional environment could then raise the possibility of a physical supply shortfall when the deliverability surplus comes to an end. This in turn could lead to a rapid increase in demand for fuel oil from industrial consumers, with corresponding impact on world oil markets.

In the medium term, concern about supply prospects remains very limited as long as there remain both good prospects for imports from Canada and substantial volumes of gas "behind the pipe" in the United States. In the longer term, the effects of the changing environment on contracting practices will require natural gas purchasers to pay the actual market price of gas. With prices tied to market conditions, purchasers should be assured of receiving adequate supplies. However, the prices they pay for those supplies will no longer be at the artificially low levels that prevailed under contracts with fixed prices entered into during the era of full wellhead price controls.

(iii) Japan

In Japan, production interests are not significant for the future of the industry because domestic supply prospects are limited. Natural gas supply is provided by the linked operations of major consumers (electric power companies, city gas companies, and one steel company) with trading houses to develop import contracts on a project-by-project basis. There has been a genuine need for long-term relationships to be built between buyers and sellers because of the financing needs and long construction lead times of an industry which is essentially built on liquefied natural gas (LNG). This structure sets the basis for the conventional analysis of the matching of future supply and demand, which tends to be conducted in the context of normative national energy planning targets (on the demand side) and the number of committed and conceived projects for LNG (on the supply side). As contracts mature and project infrastructure is depreciated, there may be room for more flexible behaviour within the established long-term relations between the overseas suppliers and the Japanese buyers.

D. Conclusions

The organisation and institutions of the gas industries in the Member countries of the IEA have an important bearing on the future evolution of gas supply. Governments influence the rate of development of indigenous reserves by their fiscal and, sometimes, depletion policies. Governments also set the parameters within which industry operations are organised, either directly through administrative control or state ownership of transmission and distribution companies, or indirectly through regulatory agencies and general economic policy. In so doing, governments influence both the degree of risk and the potential rewards which may be associated with new gas supply projects. Risk exposure may also be influenced by co-operation between private or public companies in the gas industry, especially where international trade is concerned. In Europe and Japan, this form of co-operation is particularly important. To the extent that there is no abuse of monopoly power, such co-operation may be encouraged, especially where it is designed to serve consumers' long-term interests in security of supply. In the United States, the changes which are taking place in the gas industry increase the level of risk commensurate with increasing market responsiveness. Some attention needs to be given to this in the light of the future gas supply and demand balance and the way in which that balance might affect other energy markets, particularly the oil market.

IV. DEMAND PROSPECTS TO 2010

Policy making requires a set of general expectations about future developments in the supply and use of natural gas. Such projections are always subject to major uncertainties. These uncertainties are particularly acute at the present time when it is unclear whether recent developments in the oil market will lead to a prolonged readjustment of oil and energy prices. This chapter accordingly concentrates on a qualitative assessment of the underlying factors affecting gas demand. For illustrative purposes, two projections of gas demand to 2010, based on alternative sets of assumptions about the develoment of oil prices and of gross domestic product (GDP) are shown in Appendices I and II. The results of these illustrative examples and their implications for the role of natural gas in total energy supply are summarised in section C below. Sensible interpretation of the analysis should bear in mind the institutional framework described in Chapter III above.

A. Sectoral Demand Trends

The analysis of sectoral demand trends in each major country of the OECD has shown up some similar features in all regions. Many of these features have been taken into account in considering the outlook for natural gas demand in individual regions and countries. The sectoral discussions which follow serve as background to the analysis of regional demand presented in section B below.

(i)   Trends in Residential and Commercial Sector Gas Demand

Although residential and commercial markets show widely varying, temperature dependent and therefore unpredictable seasonal demands for gas, because of the large element of space-heating demand, they offer secure sales outlets in the long term. From the point of view of the supplier and distributor, this is an advantage when compared with the vulnerability of industrial markets. However, the unit costs of delivering gas to these markets can be very much higher than the costs of delivering gas to industrial users. The supply system must be designed with sufficient flexibility in production or in storage to meet both seasonal and daily variations in offtake, especially where large baseload delivery contracts do not allow for seasonal fluctuation. Different approaches have been adopted in various countries in order to cope with this problem.

Gas distributors in some countries in Europe, notably Austria, Belgium and the Federal Republic of Germany, followed a strategy in the 1970s of pursuing steady but unspectacular growth in residential and commercial markets, behind much more rapid growth of industrial sales which provided both the cash flow and (through the mechanism of interruptible sales contracts) the seasonal load balancing function needed. Other countries adopted a different strategy — that of promoting household sales of gas as a priority from the early stages of development of the natural gas industry. The United Kingdom's use of this approach was built upon the substantial town gas infrastructure already in place. As the natural gas supply built up, there was also a rapid growth of industrial sales. The United States has traditionally promoted efforts to reserve sufficient gas supplies to meet the foreseeable needs of its residential consumers. In response to shortages induced by wellhead price control, the United States imposed a priority system of curtailments favouring residential customers, and prohibited certain uses of gas by large volume consumers. Industrial consumption was also discouraged through the application of incremental pricing and other rate-making practices that have attempted to force industrial consumers to bear some of the costs of residential service. The recent development of strong gas-on-oil and, to a lesser extent, gas-on-gas, competition has reduced the viability of this strategy and is leading the United States authorities to adjust gas regulations to accommodate an increasingly competitive natural gas market. The ban on certain uses of gas by large volume customers is, however, still in force.

The expected increasing share of the residential and commercial sectors in total natural gas demand will require load balancing investments in storage capacity in some countries. Future demand growth in this sector is unlikely to be constrained by deliberate gas company strategies of promoting industrial sales in order to ensure balanced load. The major influences on demand will be inter-fuel price competition, fuel-saving new technology, and various demographic factors such as population age structures and the rate of household formation.

Distillate heating oils are still the main fuels used to heat homes and offices in Europe and Japan, and are still the main alternative fuels to natural gas in major consuming regions in the United States and Canada. Electricity is an important source of heating in areas where hydroelectric power is abundant, especially in Canada and Norway. Electricity is likely to become an increasing source of competition with gas in countries with high shares of nuclear power (Belgium and France) and also in some segments of the market in Europe and certain regions of the United States, where generally lower installation costs for electric heating may compensate for the higher energy price.

The forecasts for this sector of gas demand incorporate these general factors, together with assumptions of rising real incomes with GDP growth. Conservation in this sector results from improved equipment efficiencies, in particular because of the turn-over in the stock of appliances, and, in the later years of the period forecast, from rising real energy prices.

(ii) Trends in Industrial Sector Gas Demand

The industrial sector is the most difficult sector of gas consumption to analyse, because of the complexity of the energy substitution economics in the various processes in which energy is used. The IEA Secretariat has conducted research into this question, including the commissioning of a series of interviews with industrial fuel-users in nine industrial sectors in six European countries[1].

---

1. Belgium, the Federal Republic of Germany, Italy, the Netherlands, Spain and the United Kingdom.

This work has provided background information for the forecasts of industrial sector gas demand. Several major conclusions emerge from it:

- Industry in the mid-1980s shows a higher degree of sophistication and understanding of energy cost management than was present before the oil price increases of 1979-80. The results of this learning process are unlikely to be undone if oil and energy prices fall and remain at low levels for a relatively long period of time.

- Where energy costs exceed a minimum level of manufacturing costs (excluding raw material costs and depreciation) then investment options for conservation or fuel switching tend to be actively sought. When energy costs are less than about 10%, interest in this area is often considerably diminished. If oil and energy prices remain lower in the late 1980s and 1990s than their average level in the early 1980s, then more companies will obviously fall into the category of those with "diminished interest".

- Required pay-back periods for energy investments seem to have lengthened for those industries whose cash flow position has improved in the mid-1980s. Investment would tend now to be undertaken for projects giving pay-back periods of three years and longer compared with two years or less at the depth of the early 1980s recession. In the more energy-intensive industries, pay-back periods of up to four or even five years are reported.

- In the raising of steam for industrial use there is a definite sense of three-way competition between natural gas, fuel oil and coal. The lengthening of acceptable pay-back periods is working to the advantage of lower priced but higher capital cost coal.

- Behavioural changes and minimal cost energy conservation measures have accounted for most of the savings made in specific energy consumption in the less energy-intensive sectors since 1980. However, heightened investment capability and awareness of conservation possibilities are likely to tend to continue to promote further savings, even in an environment of stable or declining real energy prices. A substantial and sustained decline in energy prices, however, could reduce incentives.

In general, for industrial fuel substitution a tentative conclusion might be that the variable fuel cost advantages of coal versus fuel oil and gas will offset additional capital costs in today's investment climate for a user who fulfills three broad requirements:

- load factors of 5 500-6 000 hours per year (i.e. at least two shift operations);
- annual fuel needs of 40 000-50 000 metric tons of coal or coal-equivalent (approximately 30-35 million cubic metres of gas, or 600-700 barrels per day of fuel oil).
- sufficient space to store coal.

The age of existing boilers and the costs of complying with environmental requirements also play a role in decisions for conversion. If the variable fuel cost advantage of coal is no longer expected to exist over a period of time, then coal remains clearly at a disadvantage for all users because of the higher capital and maintenance costs associated with coal-fired installations compared with oil- and/or gas-fired facilities.

Natural gas is used directly in various industrial processes and indirectly as a boiler fuel for steam-raising. It is important to distinguish these two classifications when assessing the prospects for demand. A third use, mainly in the general engineering sector, is space-heating, where the economics governing substitution of fuels or conservation may be analogous to those governing commercial uses of gas.

In some areas of process use, gas has been almost completely withdrawn as a fuel, as a result of its uncompetitiveness with respect to coal or coke. In this category can be counted blast-furnace injection in steel mills, clinker manufacture in cement works, and certain lime-kiln operations. There are however several industrial processes where gas remains and is expected to remain competitive. This is expected to be the case in the *iron and steel, non-ferrous metals,* and *brick and ceramics* sectors. In steel manufacture, the annealing process in non-integrated plants (where no off-gas recovery is possible) is essentially dependent on natural gas for product quality control. Non-integrated plants are growing in relative importance in the steel industry in Europe. Outside aluminium (which is almost wholly dependent on electricity), the manufacture of copper and its by-products is the most important part of the *non-ferrous metals* sector in terms of energy consumption. Depending on local clay conditions, gas may have intrinsic advantages for kiln firing in the *brick, tile and ceramic industries,* and its value in glazing is matched against LPG or electricity.

The prospects for natural gas in these sectors of industry will depend strongly on the output and energy consumption prospects of the industries themselves. Demand for the fuel is unlikely to be constrained by inter-fuel competition with the range of price relationships to oil products assumed in this study. In one area of process use, however, glass manufacturing, natural gas may be vulnerable to inter-fuel competition on the forecast price-relationships. In the melting process, which absorbs the bulk of the energy used in glass manufacture (finishing accounts for only 1 or 2%) radiant heat is an essential ingredient, and the radiant heat of fuel oil exceeds that of natural gas. This technical disadvantage can in some cases offset the other advantages of gas — controllability, cleanliness and ease of handling. In Japan, the development of gas-fired equipment with radiant heat capabilities may go some way towards overcoming this disadvantage.

Natural gas is used for steam-raising in a wide range of industrial sectors. About 25% of industrial gas is burned in large boilers to generate steam. Many of these boilers can easily burn alternative fuels. Among the largest users of gas for this purpose are the *chemicals* and *paper and pulp* industries. Operations in these sectors are typically large in scale and involve high load factors. They are also energy-intensive. In these two sectors, therefore, coal is most competitive with both fuel oil and natural gas. The prospects for gas demand in large boilers in these sectors are therefore poor in any scenario in which gas prices remain linked on a calorific basis to fuel oil prices, unless fuel oil prices themselves remain competitive with coal in the long run. Steam-raising in these sectors account for 20-25% of natural gas consumption in Europe. In the United States, industrial consumption has accounted for between 30 and 40% of all end-use consumption of natural gas.

The less energy-intensive, typically smaller operation of industries in the *food processing, textiles* and general *engineering* sectors, are growing in relative importance in their contribution to overall industrial energy demand. In these industries, gas appears likely to maintain its competitiveness. Ease of handling and lack of working capital require-ments are the most oft-cited characteristics which make gas a preferred fuel if priced on a calorific equivalent basis to fuel oil. In the textile industry and some food processing uses, clean-burning may add a small extra amount to the attractiveness of gas. These less energy-intensive industrial sectors account for about one-third of industrial natural gas demand in Europe, slightly more in North America, and over half in Japan, and the prospects for future demand growth are good.

(iii) Trizends in Gas Demand for Electric Power Generation

The use of gas for electricity generation will to a large extent depend on how additional coal-fired and nuclear capacities match the development of electricity demand in countries or areas where gas is a baseload fuel for power generation, such as in the Netherlands, Italy and parts of the United States South Central Region (notably Texas, Louisiana, Oklahoma and New Mexico), and where there are active plans for developing coal or nuclear powered installations. In Japan, natural gas burnt in power plants continues to form part of the national strategy for diminishing dependence on oil; already here there has been a tendency for gas-fired capacity to move towards a more middle load mode of operation, and this is expected to continue. Texas, Louisiana, the Netherlands and Japan account for more than one half of all the natural gas used in this sector in the OECD.

Elsewhere gas is used increasingly for peak load operations only, in competition with oil rather than with coal or nuclear power. Baseload use of gas has declined strongly since the mid 1970s in Belgium, the Northeastern United States and, since 1979, in Germany, which formerly were significant consumers in this sector. It is not expected that this sector could again become a major consumer of gas for baseload uses in these areas and countries.

(iv) New Areas of Demand

New technologies introduce the possibility of whole new areas of gas consumption. In the history of the gas industry, technical change has from time to time led to important shifts in the main sectors of consumption — the most obvious of which have been the changes in the principal uses of gas from lighting to cooking and from cooking to heating. Over a 25 year period, such as the one covered in this study, it is possible that some new use of natural gas, based on a novel technical development, could have an appreciable impact on the demand for gas. It is difficult or impossible to predict with any degree of certainty how changing economic, technical and social needs might evolve to create new areas of demand. It is possible, however, to identify areas of current research and technical interest which may offer some potential for the future.

Four technologies may be of primary interest over the next twenty five years — fuel cells, combined cycle power generation, compressed natural gas (CNG) as a vehicle fuel and gas-to-gasoline or gas-to-middle distillates conversion processes. Fuel cells and combined cycle power generation offer efficient and environmentally attractive means of producing electricity. CNG and gas-to-synthesis fuels conversion processes may offer natural gas a significant niche in the transportation sector energy demand in some countries.

Fuel cell technology is, in the mid 1980s, still at the research and demonstration stage. In simple terms, a fuel cell can use natural gas as a feedstock to generate electricity (without burning the gas) by a sort of electrolysis-in-reverse; the hydrogen in natural gas is electrochemically combined with oxygen to yield electricity and usable heat. The advantages of fuel cells include an ability to run up and down to follow load requirements, cleanliness, and energy efficiencies of about 50% (compared with about 40% for state-of-the-art conventional power generating). Research work is continuing in Germany, Japan and the United States, and performance is considered by some major engineering manufacturers to be sufficiently reliable to warrant engineering development for utility sized cells. Small-scale, on-site use of fuel cells by industrial or commercial users is considered the most likely path for development.

Combined cycle power generation using natural gas is at a much more advanced stage of development and commercialisation. A gas-driven turbine can be combined with a steam-driven turbine using heat recovered from the flue gases to give conversion efficiency gains in the generation of electricity. This may prove economically attractive to quite a wide range of users. In the United States, gas in combined cycle use has become an attractive fuel for industrial co-generation; the economics depend partly on relative gas and electricity prices to industry, but also on the level of prices paid for surplus power delivered to the public grid by co-generating plant. As with all new technologies, it is difficult to quantify how widely combined cycle will spread and to what extent, if at all, it may cause gas to displace other fuels in the power generation sector.

Compressed natural gas as a vehicle fuel is not a new technology — it was established in Italy in the 1930s and CNG vehicles are in operation in Italy and New Zealand as well as in experimental fleets in the Netherlands, Canada and the United States. Worldwide there are between 300 000 and 400 000 natural gas vehicles powered by CNG. The economic attractiveness of CNG as a vehicle fuel is mainly dependent on relative fuel prices and fuel taxation policies and it is these rather than technical innovations which will probably determine the ultimate size of the market. Substantial growth is not anticipated under existing taxation policies and with the fuel price assumptions used in this study.

Production of synthetic liquid fuels (gasoline, kerosene or gas oil) from natural gas is now technically feasible under processes patented by two major multinational oil companies. In New Zealand, gas-to-gasoline conversion uses some of the output of the Maui field. There are no short-term prospects for further commercialisation of these technologies in the OECD world, but there is a proposal for a gas-to-gasoil conversion project in Malaysia. In general, these technologies are likely to be of most interest as a means of obtaining value from remote gas reserves which may otherwise lie dormant. They may be even more attractive where the foreign exchange component of petroleum-based fuels is high. In most OECD countries, with the clear exception of New Zealand, the relative proximity of well developed or potential markets to gas reserves means that alternative outlets are readily available for gas. This may limit the potential for projects based on these technologies in the OECD area. They may represent an interesting opportunity for both resource development and technology transfer, however, in the developing world.

B. Factors Affecting Regional Demand

(i) Outlook for European OECD Countries

Gas sales in the Benelux countries, France, Italy, Germany, and the United Kingdom account for 95% of total natural gas consumption in Western Europe, and the business represents nearly 2% of these countries' combined GDP. Natural gas is a widely traded commodity — about half of the annual quantity sold in Europe is traded internationally. The extension, or even mere maintenance, of the contribution made by gas to energy supply in Europe and countries of the OECD will depend on a continuing high level of trade in gas. Table IV-1 (page 59) shows IEA Secretariat estimates of possible future gas demand in Western Europe as a whole. A sectoral breakdown of the figures is given in Appendix II, Table 1.

Other estimates of possible future West European gas demand, including those submitted by Member governments in the course of the IEA's annual review of energy policies and programmes, and work on the European Community countries by the EEC Commission, indicate aggregate figures towards the bottom end of the range shown in this study for 1990 and lower figures for the year 2000. Apart from different assumptions on GDP growth and oil and gas prices that underlie these estimates, two main explanations for the lower alternative figures are:

- greater optimism about the pace of improvement in energy efficiency in some countries; and
- the implementation of agreed strategies towards minimising gas use in power generation, where applicable.

All these estimates were made before the sharp fall in oil prices experienced in the first months of 1986, which may tend to diminish incentives to energy efficiency and investment in non-hydrocarbon electricity generation.

(a) Residential and Commercial Demand

The residential and commercial sector is the largest single sector of gas consumption, and accounts for about 50% of European sales. Consumption in this sector totalled over 100 bcm in 1984, which is the energy equivalent of about 1.8 million barrels per day (mbd) of oil. For comparison, about 2.5 mbd of oil products are consumed in Europe by residential and commercial consumers. The relative importance of residential/commercial demand in overall demand for natural gas is expected to continue to increase. Gas can probably remain price-competitive with oil products in this sector, even with the profile of falling real oil prices to 1990 which is assumed in this study. Natural gas is expected therefore to continue to gain market share in this sector at the expense of oil.

The prospects for consumption growth vary from country to country. In the Netherlands the residential/commercial market can be considered to be mature in the sense that a very large share of households and other users are linked to the gas grid. In the United Kingdom there is still scope for increase in gas consumption as installation of central heating becomes more widespread. In Austria, Germany and Italy the gas distribution grids are being expanded in many regions and the number of customers is growing. Several minor markets are not yet fully developed. These include the markets in Denmark, Ireland, Finland, Spain and

Sweden. In Greece, Portugal and Turkey, it is planned to introduce gas in this sector before the end of the century.

(b) Industrial Demand

In 1983, consumption in the industrial sector was 71 bcm; in 1984, a year of recovery and of especially strong demand from the chemical industry, consumption increased by about 10% to around 77 bcm. In both years, this sector accounted for about 35% of Western Europe's total consumption of natural gas. The share of industrial sector gas demand in the total has fallen from 38% in 1980, which was the year of peak industrial sales, and from 41-42% in the early 1970s. This decline in the relative share of industrial gas demand reflects the normal trend as gas markets develop.

The relative importance of industrial sector demand in total demand for natural gas is expected to continue to decline, largely as a result of the steady growth of the residential and commercial sector. Allowing for cyclical effects which boosted consumption in industry in 1984, no major increase in the absolute volume of industrial sales is expected by 1990. Some growth in this sector is forecast for countries where the national transmission grid continues to be expanded (notably Italy, Spain and some regions of Germany) but this will probably be offset by declining sales for steam-raising in other countries, as coal increases its market share among large industrial fuel users. A period of very weak oil prices in the late 1980s could inhibit the expansion of coal use, but at the same time this would tend to make it more difficult for gas to compete with heavy fuel oil in the (limited) sector of the industrial market where customers have dual-fuel, oil and gas facilities.

In the late 1990s, the market share of natural gas is forecast to increase more rapidly in the high oil price case than in the low oil price case; however, lower total energy consumption in European industry — which is in turn a function both of lower output resulting from a weaker economy and increased price induced conservation — implies that the absolute volume of gas consumed in industry could be lower with higher oil prices than it would be in the low oil price case.

(c) Power Generation Demand

Natural gas has had a relatively limited role in the power generation sector in Europe, as compared with the United States or Japan. Seven

European countries have used domestic natural gas to generate electric power — initially because alternative, higher value outlets were not available. As major coal-fired and nuclear plants have come on stream in recent years there has been a significant falling off in the use of natural gas for baseload electricity generation. The situation differs between countries:

-   In *Austria,* natural gas is an important component of fuel supply to the electricity industry, for environmental reasons in the Vienna region, but also because of its flexibility as an additional fuel in combination with coal in years of low rainfall.

-   In *Ireland* economic exploitation of the Kinsale Head gas field has been made possible by the even demand load provided by sales of gas for electricity generation (and fertilizer production). In the medium term, as demand builds up in residential and commercial heating markets, the electricity sector is likely to continue to provide a valuable load-balancing function.

In both these countries natural gas is likely to continue to have an important role in power generation. The absolute volumes involved are small.

Three other gas-producing countries — the Federal Republic of Germany, Italy and the Netherlands — are likely to account for the bulk of gas used in power generation:

-   In *Germany* gas has been gradually withdrawn over the last ten years from baseload generation. An increasing number of nuclear plants has come on stream and German-produced coal has a guaranteed outlet in the electricity sector. Gas is unlikely to recapture that market, even with falling oil prices in the late 1980s and in the low price, high energy demand growth scenario. Gas use could be continued in peak load and seasonal operations.

-   In *Italy,* the use of gas in power generation has increased — in part to satisfy take-or-pay obligations in the Algerian import contract, and in part because the alternative fuel in baseload use, at least in the short term, is fuel oil. In particular this has been necessary in the build-up phase of all major import contracts, Dutch, Soviet and Algerian. The use of oil or gas for baseload generation is likely to continue to 1990, and to a smaller degree to 2000. This conclusion will be reinforced if the general level of oil and gas prices in the late 1980s remains below their early 1980s level.

— Temporary additional sales of gas for baseload generation of electricity in the *Netherlands* are scheduled to be phased out in 1987. Government policy is to expand the use of coal and nuclear power to meet the baseload needs of the electrcity sector so that the use of gas can be phased down. The economic attractiveness for the electric utility of continuing to burn gas on a large scale as in recent years (8 bcm yearly) would, however, be greatly enhanced if oil and gas prices remain substantially lower than their average level in the first half of the 1980s.

In *Belgium* and *Spain* in recent years there has been a gradual withdrawal of natural gas from power generation. This trend is expected to continue, notably in Belgium where major investments in nuclear power will be fully on stream and are likely to provide ample capacity in the 1990s. The development of alternative industrial markets for gas in Spain has lessened the importance of electric power generation as a sales outlet for gas. In both these countries, the high cost of imported gas has militated against its use in power generation. In *Denmark*, however, there are contracts to supply small quantities of gas to power stations for a limited period. The amount will depend on the development of residential and industrial sales.

In general, the most significant conclusion for this sector is that as a result of the progress of coal and nuclear power, there is unlikely to be a recovery in gas consumption to the levels seen even in the mid-1970s in this sector. As long as government support for coal- and nuclear-based electricity remains firm in the face of lower oil and gas prices, this conclusion is likely to hold true.

(ii) Outlook for North America

(I) United States

In some regions of the United States, the distribution network for gas continues to expand although there is little activity in the transmission sector following the completion in recent years of the Northern Border and Trailblazer systems, which were designed originally as the receiving pipeline systems in the lower 48 States for projected deliveries of Alaskan gas. One or two proposals for new transmission lines to bring low-cost gas deposits to consuming regions have been made, such as pipelines to serve enhanced oil recovery projects in California. In general, however, the transmission and distribution grid will not be a constraint on the future evolution of the market for natural gas. As much as 620 bcm of gas has been moved by the system, against present sales of less than 500 bcm. It has been estimated that there is capacity to move up to 850 bcm, although of course that capacity does not match exactly at any given time the sources of available supply with the centres of potential demand. Given the existing infrastructure, the key determinants of incremental future natural gas demand will be the economics of switching between fuels, without the need for cost-recovery in additional grid investment.

Demand is expected to recover further, following 6 to 7% growth between 1983 and 1984, in a short- to medium-term environment of continuing surplus deliverability and falling prices. Medium-term growth in gas demand is forecast, consistent with continued economic expansion, and growing energy demand as a result of falling real oil prices. Forecasts for the United States are made independently from the analysis of supply prospects[1]. In particular, the higher end of the demand range may be unrealistic given constraints on the potential for growth in supplies in the long term. A fuller discussion of the supply prospects is given below in Section V.D.

(a) Residential and Commercial Demand

Residential and commercial consumption of gas in the United States is concentrated in seven states of the northeast and midwest (New York, Pennsylvania, Ohio, Illinois, Indiana, Michigan and Wisconsin) and in Texas and California. The prospects for demand vary regionally (see below) but relatively stable demand for gas is expected in this sector to 2000; demand could be stimulated over a period of ten to fifteen years by a lower level of oil and gas prices. Recent evidence compiled by the United States Department of Energy and the American Gas Association suggests that demand for gas by residential users responds fairly strongly over the long run to changes in prices. Commercial sector gas demand is believed to be even more price-sensitive.

---

1.   Appendix II, Table 2 shows the forecasts for demand in each major sector of consumption.

Regional variations in demand prospects in this sector may be quite large. No growth in demand is expected in the mature and important markets of Illinois and the principal midwestern consuming states. Any change in sales volume is likely to be downwards as efficiency in use improves with the turnover of the boiler stock. Companies which have traditionally served these markets are adopting strategies of diversification into other regions. Partly as a result of these strategies, in other areas, notably New England and the Pacific northwest, demand is likely to increase. In New England, 50% of residential heating is still oil-based, although many consumers whose principal heating fuel is oil are already connected to a gas supply for cooking purposes. As customers replace aging oil-fired equipment, gas is expected to increase its market share. In the Pacific northwest, where electricity has traditionally had a large (by national average standards) share of the heating market, anticipated increases in relative electricity prices may create some competitive advantage for gas.

On a national level, natural gas continues to gain market share in this sector from other fuels. Conversions from other fuels to gas outnumber conversions away from gas. Forty-three per cent of single-family homes in the United States are currently heated by gas, but between 45 and 50% of new single-family units are supplied with gas for space-heating every year. Against this trend to a slowly rising share of the total energy market in the sector, some increases are expected in the efficiency of boilers, so that in volume terms, the net increase in demand for gas is likely to be offset as old boilers are replaced. The impact of this technological change will not be affected by a lower general level of energy prices, as the efficiency improvements are already incorporated into new vintages of boiler design.

(b) Industrial Demand

The economics of inter-fuel substitution in the industrial sector in the United States tends to be more subject to short-term influences and direct fuel price comparisons than in Europe. Dual-fuel facilities are more common in the United States, partly as a result of the actual experience of unanticipated curtailments in gas supply during the 1970s, from which European industry has never suffered. If the relative prices of fuels change, significant effects on the consumption of natural gas can therefore be observed without waiting for the consequences of major investment decisions to work through. Wide variations in the level of demand from year to year can therefore be expected, and forecasts must be interpreted as outline sketches of possible trend values.

The immediate and medium-term prospects for gas in this sector are for continuing recovery of demand as a result of economic growth and restored price-competitiveness of gas compared with residual fuel oil to many industrial consumers. Gas is expected to gain market share at the expense of fuel oil, but demand by 1990 will depend principally on the growth in the economy and on conservation as determinants of the overall level of energy demand. The Fuel Use Act of 1977 (which restricts the use of natural gas in certain large industrial boilers) is assumed to continue in force, but with exemptions given when appropriate according to present practice.

Approximately one-third of industrial natural gas sales take place directly between gas producers and purchasers in the chemicals and petroleum refining industries, mainly located in the major gas-producing states, such as Texas and Louisiana. The remaining two-thirds move through pipelines, subject to state and federal regulation. Regulatory developments, and the ability of the natural gas industry to develop competitive practices to maintain and expand sales in the face of gas availability which exceeds demand in the 1980s, will be critical influences on the future prospects for this two-thirds of sales. No changes from existing regulatory conditions are assumed in our analysis, so that most industrial consumers are assumed to continue to be able to have access to gas at prices which can compete with residual fuel oil.

In the critical chemical industry, which accounts for up to a quarter of demand, gas consumption is expected to grow slowly over time with increases in output, except in ammonia and methanol manufacture, where the trend demand is expected to be relatively flat in view of the world prospects for these industries. In other industries which consume gas for process use, demand is expected to grow in line with the general level of energy demand, conditional upon the output prospects of each sector. Gas demand in the important refining sector (10% of industrial demand) is likely to be stronger if there is an added impetus to oil consumption from declining real oil prices through to 1990. After 1990, gas use in refineries is projected to be in decline, as low-cost gas works its way out of the system and refiners rely on internal fuels in place of purchased gas which would have a higher alternative value. This is shown in the Energy Sector in Appendix II Table 2. The Energy Sector also includes gas used for steam-raising in the enhanced recovery of oil, whose growth prospects could offset the decline in refining use — especially if high oil prices cause enhanced recovery to become increasingly economically attractive. Pipeline fuel and loss is, in this calculation, a constant 3.3% of total demand.

For the use of gas for raising steam, the principal competition is expected to continue to be residual fuel oil for all except very large boilers. This conclusion is primarily a function of the oil price profile assumed in this analysis, under which the marginal boiler-size at which coal can be attractive as an alternative fuel, requiring investment, is likely to rise over time (at least until such time as the oil price rises again compared to coal). The prospects for gas then depend on two-way competition with residual fuel oil for a sizeable part of the market, and gas demand growth will be conditional on the availability and cost of supplies.

Assuming sufficient supply availability, development in industrial gas use from 1990 to 2000 is expected to be relatively stable, as the intensity of gas use per unit of industrial output is forecast to decline:

- if energy prices are high because of further price-induced conservation;

- in the case of higher growth and lower prices, because of high levels of profitability and investment and a fast rate of technological innovation.

These factors will offset the effects on demand of growth in industrial output, so that the range of industrial gas demand in 2000 could fall between 150 and 190 bcm. This compares with consumption of 175 bcm in 1984 and 225 bcm in 1972.

(c) Power Generation Demand

Almost four-fifths of gas in power generation is used in four States — California, Louisiana, Oklahoma and Texas. For environmental reasons, gas consumption in other regions may continue to be attractive, and in some areas the prospects for natural gas demand in this sector are good. Over the very long term "select gas use" (the combination of gas with another fuel in the same installation in order to meet emissions standards) could prove an increasingly attractive option. However, in terms of the absolute volume of demand, developments in the four main States will be overwhelmingly important. In these States, the rate of completion of coal-fired and nuclear power stations is expected to be higher than the national average, so that gas is unlikely to retain its share of the power generation market. Coal transportation costs will be a key variable; if the prospect is for them to decline further, then coal could considerably enhance its competitiveness in these key regions. The Fuel Use Act is assumed to continue to act as a constraint in the development of new gas deposits exclusively for the generation of electric power.

(II) *Canada*

A major programme of extension of the nation-wide gas distribution grid to more remote areas is expected to enable natural gas to increase its share of the residential and commercial sector energy market from its current level of about 38% to about 43% in 1990 and to between 44% and 48% in 2000. The ending of financial incentives designed to promote conversions to natural gas is not expected to have a major impact in reducing the rate of penetration of gas in this sector. Subsequently, demand growth may be slowed by market saturation. Gas consumption in this sector in the newly connected regions of Quebec is likely to be modest because of competition from electricity based on hydro power. Eighty per cent of new homes in Quebec are constructed with electric heating, and the ratio has not moved significantly in favour of gas in recent years despite increased availability of supply[1].

The industrial sector, which currently accounts for just over 40% of natural gas demand is expected to show strong growth with the development of distribution networks and the need to balance load factors in the expansion phase. The main influence on gas consumption will be the outlook for continuing growth in energy demand in industry. But a substantial part of the growth of gas is expected to come from industrial and mining plant conversions from other fuels in areas which do not currently have access to gas supplies. Industrial demand could also be supported by heavy oil upgrading projects, steam-raising for enhanced oil recovery and miscible gas re-injection in the oil production sector in Alberta. This will tend to boost demand in the event that oil prices are relatively high so that such activities would tend to be economically more attractive. The forecast range for industrial sector gas demand is thus narrower than for residential and commercial sector

---

1.   Table 3 in Appendix II shows the outlook for demand in this and other sectors.

demand, since the depressing effect of high prices on demand in conventional industries would be compensated by higher demand in non-conventional oil recovery and upgrading operations.

Natural gas is used on only a very small scale in power generation in Canada (less than 2 bcm per annum) and no change is likely given the abundance of hydroelectric, coal and increasingly nuclear power supply in the country. If gas fields offshore the Atlantic Provinces were to be developed, however, it could add to gas use in this sector. In view of the medium-term prospects for hydrocarbon prices and investment, this is not shown in the outlook in this study.

(iii) Outlook for OECD Pacific Countries

(1) Japan[1]

The use of natural gas for electricity generation will continue to be the most important market for gas in Japan. There is no national or unified regional grid for the distribution of natural gas in Japan. Fourteen qualities of reformed and manufactured gas are distributed by 248 local gas companies, and the major use of natural gas is as a feedstock to city gas works, rather than as a directly piped fuel as in other parts of the world. Residential, commercial, and industrial demand for natural gas is composed of city gas, manufactured from natural gas plus relatively small quantities of direct natural gas use. City gas produced from other feedstocks, primarily oil products, is not included in this analysis. Total gas demand in these sectors would be somewhat larger. LNG is likely to become an increasingly important feedstock for city gas manufacture, rising from 55% in 1983 to up to 70% to 75% by 1990 and to a maximum of about 75 to 80% by 2000. In both residential/commercial and industrial markets, the rate of growth of demand for natural gas will therefore be faster than the rate of growth of demand for all city gas. The upper limit to the use of LNG as feedstock may be set for one reason by the rigidity of LNG supply: constant shipment of LNG improves the supply economics of an LNG project, but this does not meet the seasonal demand fluctuation and LNG must therefore be supplemented by other fuels (naphtha, LPG or coal-derived gases) to meet demand at its peaks.

---

1. The forecast demand for natural gas by sector of consumption in Japan is given in Appendix II, Table 4.

(a) Residential and Commercial Demand

The residential and commercial sector accounts for a relatively small share of total natural gas consumption in Japan — about 19% in 1983. This is despite the very rapid growth in the number of gas customers in the household sector in the last 15 years, and the high proportion of households which are connected to the regional grids of the three major gas companies, Tokyo Gas, Osaka Gas, and Toho Gas. Gas is principally used for water-heating and for cooking, and has only a minor space-heating role because of the mild climate in the major consuming areas (the coastal strip from Tokyo to Hiroshima), together with the cost advantage of kerosene over city gas. Average consumption of gas per household connection is therefore low by international standards, at about 4 000 megacalories (approximately 16 Million British Thermal Units (MBTU)) per annum compared with about 13 000 megacalories (50 MBTU) in the United Kingdom and 30 000 megacalories (120 MBTU) in the United States. Prospects for future expansion of natural gas use in this sector will depend on increasing use of LNG as a feedstock, at the expense of naphtha, LPG, and coal, as well as on direct growth in the end-use residential market for gas where it competes with other oil products such as kerosene, light heating oil and LPG.

The capacity of the delivery system will not be a constraint on the expansion of the market in the foreseeable future. The three major gas companies account for about three-quarters of all gas sold to residential and commercial markets, and within their distribution areas and those of 18 smaller companies, a major programme of conversion from city gas quality to natural gas is being carried out. This applies both for city gas works feedstocks and for direct delivery of high calorific value gases (usually 11 000 kilocalories per cubic metre) by pipeline. The programme should be completed for the major areas by 1988 to 1990; its effect will be to enhance considerably the amount of energy which can be delivered by existing pipeline networks. In the Tokyo, Osaka, and Nagoya regions, delivery capacity will be at least double, in energy terms, what was possible in 1983, with no major additional investment in pipeline infrastructure.

The rate of growth of commercial sector gas demand could be very much faster (at around 4% per annum or higher to 1990, and 5 to 7% per annum from 1990 to 2000) than the rate of growth of household gas demand. This expectation is based largely on the prospects for growth in the use of air conditioning and co-generation, where a special rate structure and new technologies enable gas to be competitive with electricity.

In the case of high oil prices, natural gas would tend to increase its share of the energy market relative to oil, at a faster rate than in the event of the lower oil prices. However, the growth in total energy demand is limited by conservation and the development of renewable energy sources. If demand for natural gas in the residential and commercial sector were to grow so that annual demand exceeded 10 to 12 bcm, then a limited amount of incremental investment (one major new trunkline in the Tokyo area) would be required to permit the expansion foreseen in the later years. No new investment in major distribution grids would be required for a lower growth case, and development of a national transmission network and major institutional changes is in any case unlikely to be necessary.

(b) Industrial Demand

Natural gas for the industrial sector in Japan accounts for about 2% of industrial energy use and for about 10% of natural gas consumption. It is supplied in three ways: by direct supply of pipeline gas from domestic fields such as Niigata to local industries, by direct import of LNG and by pipeline distribution of reformed gas from the city gas companies for whom LNG is now the major feedstock. The available supply of indigenous gas is expected to increase and industrial consumption of gas from this source could grow accordingly.

Industrial demand for city gas is concentrated mainly in the machinery, general engineering, and food processing industries in Japan, which are generally not very energy-intensive and thus evaluate city gas as a premium fuel because of its ease of handling and lack of storage needs. This contrasts strongly with Europe and, especially, the United States, where gas is widely used in bulk steam-raising as well as process uses in the more energy-intensive industries (chemicals, non-ferrous metals, paper and pulp). This accounts for the relatively low volume of city gas use in Japanese industry. An important factor in future expansion of gas sales in this sector as in the residential/commercial sectors will be the doubling of capacity and lowering of unit transport and storage costs as the major gas companies complete their programmes of conversion to high calorific value from low calorific value gas.

Growth in demand in the industrial sector is forecast to result both from strong increases in the direct supply of indigenous natural gas to industry and from expansion of sales of distributed city gas. The increasing share of LNG as feedstock to city gas manufacture, as described above, will also contribute to the increase in natural gas sales attributable to this sector. Industrial users are not expected to contract for new direct LNG supplies because of the inability of most industries to offer large enough individual purchases to make economies of scale attractive. The single existing direct LNG contract is assumed to continue to its expiry date in the low demand growth scenario, and to be renewed beyond that date in the high growth scenario. No other direct LNG sales are assumed in this sector.

Industrial consumption of distributed gas, which supplied slightly more than 3 bcm in 1984 (of which some 85% originated as natural gas), is anticipated to grow by up to 30 to 40% by 1990, an annual average growth rate of between 4% and 6.5%. Growth from 1990 to 2000 could be at the slower, but still substantial rate of about 4% per annum. Subsequent growth would be likely to continue at a slower rate. Direct sales of indigenous natural gas could rise from 0.8 bcm in 1984 to over 1 bcm in 1990 and to between about 2 and 3 bcm in 2000. No further major development of existing city gas company pipeline networks would be needed to achieve the volume of sales at the lower end of these ranges; limited additional investment in new trunk pipelines would be required by one or two of the major distribution companies to achieve sales of gas at the higher end of the ranges.

(c) Power Generation Demand

The main use of natural gas in Japan is in power generation, mostly in the form of imported LNG. In this form it has substituted for low sulphur fuel oil and, in some cases, crude oil and naphtha as a clean-burning fuel for power station boilers. Future consumption is planned to develop in accordance with existing supply contracts and power-station construction plans. This is likely to lead to a doubling of LNG use in this sector between 1983 and 1990. Thereafter, options remain open both with respect to the type of fuel supply planned for new power stations (which may be nuclear power, coal or LNG) and with respect to the utilisation rates of the operating LNG-fired plants. The latter factor is a particularly important influence on future demand for natural gas, and will be determined both by the evolution of total demand for electricity and by its seasonal profile. Electric utilities currently plan to increase the generating capacity of nuclear and coal, and put less emphasis on LNG. As the proportion of nuclear power rises, the role of LNG in Japanese power stations is moving away from baseload to middle load operation, and annual utilisation rates for these power plants could move to below 50% with a corresponding depressing effect on LNG demand.

The effect of such a move towards lower load factor operation for LNG-fuelled power stations could be that demand for gas in this sector would reach 33 bcm in the early 1990s, then remain at a stable average annual level of about 35 or 36 bcm to 2000 and beyond. On the other hand, continuing growth in demand for electricity, or other energy market developments, such as the lower level of hydrocarbon prices which seemed to be presaged by oil markets in early 1986, which would make baseload operation of LNG-fuelled plants attractive, could lead to a level of about 39 bcm of natural gas demand in this sector by 1990, rising to up to 45 bcm by 2000 and 2010. New import supply projects, now in the conception stage, would have to be developed in the 1990s to meet such a requirement. However, in practice the range of options open to electric utilities, both with respect to the operating modes of LNG-fired power stations, and with respect to the mix of generating capacity installed, may be narrowed considerably by actual LNG supply contracts to which they are committed and by broader strategic considerations establishing the "best mix" of capacity, particularly with respect to nuclear and coal.

(II) Australia[1]

The Australian gas market consists of a few urban communities of moderate to large size separated by long distances. In general, the location of the Australian gas fields does not correspond with the distribution of population and industry and consequently the natural gas transmission pipelines are dedicated to specific urban centres. Interconnection of some of the pipelines is likely in the future as production from currently producing gas fields declines.

The residential/commercial market for gas is centred on the capital cities of the states of Victoria, New South Wales, South Australia and Queensland. In Victoria residential markets are very well developed and

---

1.   Appendix II, Table 5 summarises the forecasts of natural gas demand for Australia.

gas is a dominant fuel for heating and cooking. Prospects for further growth in Victoria are fairly limited. In other States, there is still potential for gas to increase its market share but the possibilities are limited by the availability of competitive fuels, primarily electricity, and the dispersal of population centres which currently have no access to natural gas.

The gas demand of the industrial sector has, in the past, been more prone to growth than the residential/commercial sector. Industries are normally more easily served with gas than the residential market although in some circumstances gas must also be transported over long distances. The major industrial use of gas in Australia is in the petrochemical, alumina and aluminium industries. Industrial gas demand is projected to increase as transmission systems are extended to major industrial locations currently consuming liquid fuels, and by further backing out liquid fuels and some solid fuel in areas already served by natural gas. However, the process is projected to continue at a significantly slower rate than in the past.

Some growth in the use of gas for electricity in the near term to the early 1990s is likely because of the need in the build-up phase of the North West Shelf project to have a baseload outlet for the gas. This project is currently supplying 1.1 bcm per annum over 20 years to the Western Australian market via a 1 500 kilometre pipeline from Dampier to Perth, and will begin supply of LNG to the Japanese market in October 1989. However, in this sector in many areas gas competes with easily available and relatively cheap indigenous coal. In the long term, there is therefore no significant growth projected in the demand for gas in electricity generation.

(III) New Zealand[1]

The most important consuming sectors of gas in New Zealand are industry and electricity generation, each of which account for over 40% of total gas use. The geography and economic structure of the country imply that electricity will remain an important consumer of gas although industrial gas use could increase in relative importance.

---

1.   The outlook for demand is summarised in Appendix II, Table 6.

A special feature of the New Zealand market is the use of natural gas for the production of synthetic liquid fuels, from the gas-to-gasoline process. Production of gasoline with this process began in 1985, and by 1990 it is likely to at least equal, and possibly exceed, power generation in terms of the volume of gas used as an input. When fully operating, some 1.3 bcm of gas a year will be required to supply the gas-to-gasoline plant.

Demand in bulk industrial uses is unlikely to grow strongly, because the outlook for consumption in methanol manufacture (the dominant industrial use) is static. There may be some growth, especially in the high demand scenario, in pulp and paper and basic metals' gas demand. Most of the growth in this sector is likely to come from lighter industries.

In the residential and commercial sector, gas competes mainly with electricity, and electricity is projected to continue to dominate this market. Gas is likely to increase its current market share at the expense of solid fuels. This sector will remain, however, by international standards, a comparatively small sector of demand. With the commissioning of a new gas pipeline to Gisborne in early 1985, all the major population centres of city size in the North Island are now supplied with natural gas. Uniquely to New Zealand, the transportation sector now provides a market which is significant in terms of national consumption, about 0.1 bcm of gas in vehicles which use compressed natural gas (CNG). The number of CNG vehicles is currently over 100 000 in New Zealand. The previous Government's original objective of 150 000 CNG vehicles is unlikely to be achieved under the oil price assumptions used in this study; however, demand in this sector will probably double fairly easily given the existing impetus of the conversion campaign.

The development of the various markets for natural gas in New Zealand will depend on the performance of the Maui and Kapuni fields over their lifetimes and on evolving supply prospects, as much as on the international energy environment. Under present and foreseeable circumstances, there is no available economic option for offshore New Zealand gas in export markets.

C.   Illustrative Projections to 2010

A set of forecasting assumptions have been used in order to provide the technical parameters necessary to develop demand estimates. These represent a series of assumptions on individual aspects of the energy environment — crude oil prices, economic growth, industrial activity and relative fuel prices. They are technical parameters and do not necessarily reflect an IEA view of how these factors are likely to evolve over time; the IEA view is that such evolution is and will remain unknown. The outlook for natural gas demand is derived from assumptions made for these variables for each of the main OECD regions — Europe, North America, and the Pacific — and from analysis of the evolution of the market share of each primary and secondary fuel in total energy use in each sector of consumption. In analysing total energy demand in each sector, an output and price elasticity approach was used. In the industrial sector the energy demand analysis incorporated both industrial production, derived by regression relationships from a given country's actual and forecast GDP growth, and technological progress through the turnover of capital stock, at a rate individually assessed for each country. In the residential and commercial sectors, energy demand was determined by the evolution of the stock of buildings and of its average age, combined with per unit energy consumption trends which themselves evolve according to behavioural and structural conservation and efficiency changes. Calculation of the share of each fuel in the respective sectors of consumption involved functional relationships between demand and relative fuel prices. In the industrial sector, a distinction was made between demand for non-substitutable fuels (mainly electricity and certain process uses) and substitutable fuels (mainly the requirements of the industrial heat market). In the latter case, investment decisions were taken to be additionally determined by the share of energy costs in total production costs. Appendix I gives details of the growth and price assumptions used in this work.

A range of demand estimates for natural gas for each of the main OECD regions, Europe, North America and the Pacific has accordingly been calculated. The ranges are summarised in Table IV-1. Detailed sectoral figures are shown in Appendix II. Table IV-1 shows that total annual natural gas consumption in the OECD as a whole was running at the same level of about 800 bcm, or 700 Mtoe, in 1984 as it had been a decade earlier in 1973. During this decade, however, there were important changes in the geographical location of natural gas use. Annual consumption in North America fell by over 100 bcm, mainly as a result of a large fall in the use of gas in industry and oil refining in the United States. In Europe, consumption increased by 70 bcm, or over 50%, as large new residential, commercial and industrial markets were developed in several countries and as major transmission infrastructure was put in place to bring new supplies from Algeria, Norway and the Soviet Union. In the Pacific region, there was a fivefold increase in consumption, from 10 to over 50 bcm, as LNG import projects were completed in Japan as well as new domestic gas supply projects in Australia and New Zealand. North America accounted for over 80% of all OECD natural gas consumption in 1973; as a result of these changes in the pattern of consumption, by 1984 North America's share was two thirds, with Europe accounting for more than a quarter and the Pacific OECD countries the balance.

Table IV-1
OECD Natural Gas Demand: 1973-2010
(billion cubic metres)

	1973	1984	1990	2000	2010
Low Demand/High Oil Price					
Europe	140.3	212.0	224	248	258
North America	649.2	540.7	505	568	566
Pacific	10.5	53.9	64	72	77
Total (bcm)	800.0	806.6	793	888	901
Total (Mtoe)	696.9	702.6	691	774	785
High Demand/Low Oil Price					
Europe	140.3	212.0	244	280	305
North America	649.2	540.7	589	658	642
Pacific	10.5	53.9	75	91	98
Total (bcm)	800.0	806.6	908	1 029	1 045
Total (Mtoe)	696.9	702.6	791	896	910

Source: Energy Balances of OECD countries and IEA Secretariat.

The share of natural gas in total primary energy requirements (TPER) in OECD as a whole, which was 18.8% in 1983, is projected to decline slightly to approximately 17% by 2000 in both cases. The decline will mainly take place in the electricity sector. The share of natural gas in the electricity sector is expected to decline by three percentage points from the current level of 9.2% to about 6% in 2000. Note that this relative decline is projected even after allowing for continuing gas use in baseload generation of electric power in some key countries as a result of low oil and gas prices in the medium term. The contribution of natural gas to the industry sector is also likely to fall slightly from the 1983 level of 24.6% to 23-24% by 2000 mainly due to the interfuel substitution from gas to coal and electricity in North America. On the other hand, gas' share in the residential/commercial sector in OECD total is likely to increase from the current level of 32.8% to 33-34% by 2000.

The share of natural gas in OECD as a whole is expected to decrease a little in the coming 15 years, but the results of the projection are diverse because of regional specific conditions. For instance, in Europe, practically no change in gas' share is expected, since the gain in the residential/commercial sector may be offset by the losses in other sectors in that region. It is expected to stay at approximately 15%. On the other hand, in North America, the share of natural gas in TPER is likely to decline by more than three percentage points from the current level of 24% to 20-21% by 2000 due to the declining use of gas in the industry sector. In the OECD Pacific countries, contrary to the situation in North America, gas is predicted to increase its share in TPER especially in the coming few years because of the increased use of gas by the electricity sector. The share of natural gas in the OECD Pacific countries, which was 8.4% in 1983, is assessed to rise to 12-13% by 1990. Although gas' share in that region could decline in the 1990s, it could stay at 10-12% in 2000.

As indicated by the relatively small changes in natural gas' share, natural gas is not likely to be a major subject of interfuel substitution in the coming years given the changes in relative energy prices assumed, which are given in detail in Appendix I. The possibility of interfuel substitution between natural gas and other energy sources is mainly limited to the cases of the industrial and residential/commerical sector in Europe (between gas and various oil products/coal/electricity), the industry sector in North America (between gas and electricity/coal) and the electricity sector in Pacific (between gas and coal/nuclear).

In summary, there has been a relative decline of natural gas use in North America, a sectoral shift within overall stability in its market share in Western Europe, and relative growth of gas versus other fuels in the Pacific. These changes have occurred during a period of relatively high oil prices from 1973 to 1985. In early 1986, real dollar oil prices fell sharply towards the levels of the early 1970s. In all OECD markets there is competition between natural gas and oil products as well as between gas and other forms of energy, so that movements in oil prices translate partially or fully into movements in gas prices. Different industrial and regulatory structures, and different contracting practices, mean that there are varying time lags in different OECD countries. The pass-through of price changes can be instantaneous in some countries (indeed in parts of the United States intrastate market gas prices may lead changes in heavy fuel oil prices, as a result of gas-on-gas competition) and may take up to several months in others, such as Japan, where new import reference prices may have to be negotiated after there have been large changes in crude oil prices and currency parities.

If real oil prices were to remain at a level roughly equivalent to what they were in the early 1970s then overall energy intensity could increase and gas and oil would tend to improve their competitive position vis-à-vis coal and electricity. Energy- and gas-intensive industries such as the European fertilizer and other chemical sectors, which have appeared particularly vulnerable to competition from newer lower feedstock cost industries in the developing world, might remain economically healthy longer than expected. The relatively strong growth in total energy consumption implied in the analysis of this study, and the strength of gas consumption in sectors such as power generation and basic industrial steam raising in certain countries, reflect a perception of oil and gas prices at these lower levels in the medium-term.

On the supply side, some United States oil and gas could be shut in as uneconomic to produce in the late 1980s. In the context of surplus deliverability expected to continue in the late 1980s, this would tend to bring supply and demand into balance sooner. Some Canadian and United Kingdom North Sea gas developments might be delayed. In all cases, once the surplus of gas supplies which had been expected to be on offer in the late 1980s has been worked off, commercial gas distribution companies would then have to reconsider marketing strategies aimed at winning industrial customers. By this means, fuel oil might maintain a higher share of the total energy market in these countries than would otherwise have been the case.

V. RESERVES AND SUPPLY PROSPECTS TO 2010

A. Reserves and Supply — Stocks and Flows

Reserves of natural gas do not constitute a supply of natural gas. Reserves are stock data whereas supply is a flow towards a market demand. The supply of gas is measured as a flow which occurs at an annual, monthly, or daily rate. The notion of supply is an economic notion of flow over time. Reserves data, as stocks themselves, evolve over time, however, because of revaluation of the reserves position as well as because of depletion. The transition from reserve "stocks" to supply "flow" is by no means straightforward. This dynamic aspect is important for policy decisions at both governmental and commercial level; and serious misunderstandings can arise, and have arisen in the past, from misuse of such indicators as "reserves-life indices" or reserve-to-production ratios.

Any estimate of world natural gas reserves is fraught with difficulties. Experts differ both about how to identify reserves and about how to define them when they have been identified. Geological identification of quantities of gas, based on seismic or drilling work, is almost as much of an art as a science, involving an element of informed guesswork. Definitions of what has been identified vary from country to country, and from company to company, and involve different assumptions with respect to the degree of confidence expressed in the definitions. This chapter will attempt to clarify some of the issues involved in the assessment of natural gas reserves, and to draw conclusions regarding supply prospects.

(i) Categories of Reserves

Figure V-1 depicts the relationship of gas-committed-to-market (supply) to various categories of reserves and resources. The future supply of gas will be a function of gas committed to domestic and export markets by suppliers from reserves which can be commercially developed. Such reserves may also be termed "active" reserves. Commercial or active reserves consist of "proved" reserves, as published by various governments and other agencies, and as shown in Table V-1, but not only of proved reserves. Active reserves also include, within different confidence limits, probable and even possible reserves.

Figure V-1
Schematic Relationship of Gas Reserves and Resources

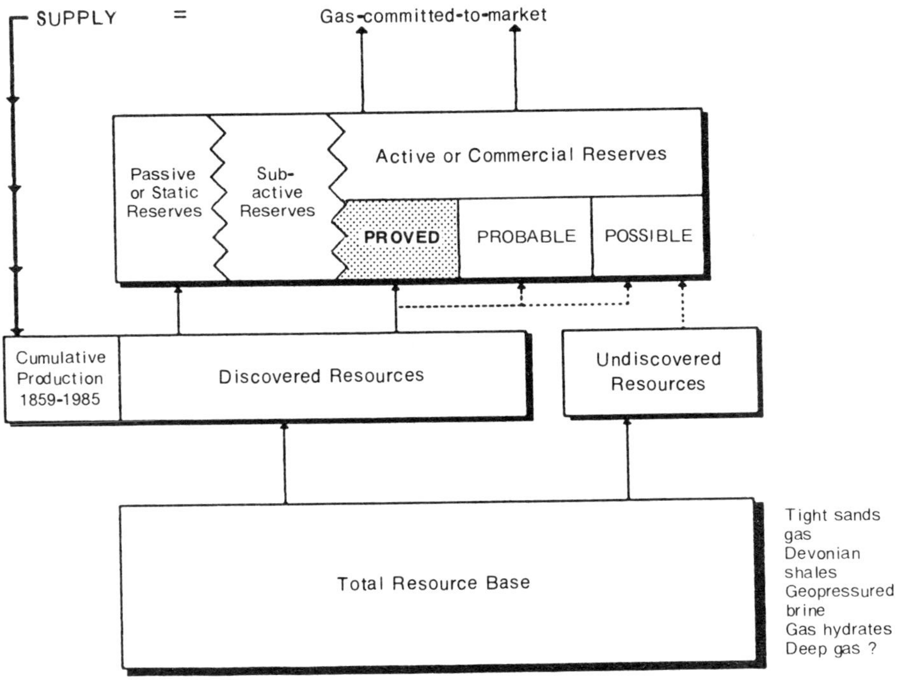

1. The shaded area "proved reserves" relates in principle to the figures shown for "proved reserves" in Table V-1. However, definitions are less certain for countries outside the OECD area, and some gas included in "proved reserves" would perhaps more appropriately be classified as passive or sub-active — notably in the Middle East and Africa, perhaps also in the Soviet Union.

Source: IEA Secretariat.

Most definitions of proved reserves use a long or short form of the following:

> "Proved reserves are the volume remaining in the ground which geological or engineering data demonstrate with reasonable certainty to be recoverable in future years from known reservoirs under existing economic and operating conditions."

Some definitions are based on "existing and expected" economic and operating conditions, although in general most companies, industry associations and governments try to narrow the definition to as technical a one as possible, abstracting from economic uncertainties. This usually means in practice an assumption of "no change" in the economic parameters (notably prices and taxes) from those prevailing at the time of the estimate.

Probable reserves are not merely reserves which might or might not exist. They are, rather, reserves which surround proved reserves — and when quantified, they are estimates of what may be produced from undrilled or incompletely evaluated portions of known reservoirs. Possible reserves can be similarly defined, but occur where there is a higher level of uncertainty regarding the characteristics of the productive zones. In any part of the world, quantification of probable and possible reserves, from which the pool of ultimately commercial or active reserves will be drawn, is difficult. Two examples may illustrate their relationship to "proved" reserves, however:

- In the United States, the quantity of gas which has actually been recovered over the years from discovered fields has consistently since 1946 been two to three times greater than the volume which was believed to be contained in the fields when they were initially discovered[1]. This has been the result of infill drilling, extensions, and the discovery of new pools in old fields. Where gas has been produced in association with oil (solution gas), enhanced recovery of oil-in-place above initial estimates has also made a contribution to higher gas production.

- One major multinational company, in calculating its available gas reserves from existing fields, adds together its proven reserves, 50% of its probable reserves and 25% of its possible

---

1. Source: "Energy Resources in an Uncertain Future", Adelman, Houghton, Kaufman and Zimmerman, Ballinger Publishing, 1983, Table 6-2.

reserves. This is considered suitable within the company's legal requirement to make a generally conservative estimate of its asset base.

In summary, the critical point is that commercially active reserves may be larger — sometimes substantially larger — than proved reserves. Proved reserves are a much narrower definition, and it is the commercial or active reserves from which future supply will ultimately flow. Proved reserves in many parts of the world are only a measure of expected cumulative output from productive capacity already in place — i.e. in the absence of future investment. Proved reserves alone are not a suitable basis, therefore, on which to make estimates of future supply.

On the other hand, not all the gas resources which have been discovered and which are known about with certainty can be considered as proved reserves. If, for reasons of distance or other reasons, readily and cheaply producible gas has no prospect of being developed and sold, then this gas must be considered a "passive" (Soviet terminology) or "static" (Exxon terminology) reserve even though its existence is known with a high degree of certainty. A large but remote gas discovery will be of less relevance when considering future supply than small, as yet undiscovered, gas pockets near to existing infrastructure. Of course, if the remote area is developed because other fields are found nearby and a pipeline is built, or if prices rise, or production taxes fall, then passive reserves may become active. In all parts of the world, there is a grey area between the passive and active categories, in which much of the debate about allocation of resources for development, risk, fiscal conditions, and potential commercial reward takes place.

Reserves, whether active or passive, proved, probable, or possible, only become reserves after investment activity (in appraisal or development wells) has converted them from being merely "discovered resources". In addition to resources which have been discovered there are also gas resources whose existence can only be speculated on, using estimates for gas contained in unexplored sedimentary basins by analogy with what has been discovered in more thoroughly explored areas. The limits of this "undiscovered" category, and of the total resource base, which is the sum of discovered and undiscovered resources, are very ill defined. First, it is often difficult to estimate the conventional undiscovered potential. Second, in addition to the conventional undiscovered resources, can be counted gas which is trapped in extremely low permeability rocks (tight gas sands), gas dissolved in geopressured brine, Devonian shales, frozen gas hydrates and, possibly, ultra-deep gas. Quantities of these are unknown, but they may multiply the total conventional resource base by factors of 50, 100, or more. Some of these resources in some parts of the world can be developed into reserves at prices little higher than today's; others could require prices which are even higher than would be required to justify large scale use of some renewable energy technologies. The relationships between all the various categories defined above are shown in Figure V-1.

(ii) Estimating Reserves

Published proved reserves will often, outside centrally planned economies, tend to be conservative assessments. In part this is because much of the raw data is aggregated from the reported reserves of private companies, which are required to make the assessments as part of their asset valuation. These companies are legally obliged not to exaggerate their asset value, and tend therefore to play safe by understating rather than overstating their reserves. In some countries, such as Norway, the authorities responsible for preparing reserve estimates may make allowances for this and other factors, and will not knowingly permit either an optimistic or pessimistic bias to appear in their figures. In other countries a more cautious approach to defining and publishing proved reserves may be taken. It is difficult to know whether assessments of proved reserves in centrally planned economies tend to be pessimistic or optimistic, but it seems likely that some gas which in the West would be classified as static reserves, not proved at all, is counted as proved gas in the Soviet Union.

As regards the total resource base, both discovered and undiscovered, two further considerations tend to cause potential resources to be understated. Firstly, natural gas occurs either in free, non-associated form, or associated with crude oil as "gas cap gas" or "solution gas". In the latter case, the gas is dissolved in the oil and separates out when the pressure of the oil drops as it is produced from the reservoir. The volume of gas recovered will depend on the volume of oil recovered. With primary production techniques, typically only one-third of the oil-in-place can be extracted. However, with enhanced recovery techniques, whose economic attractiveness depends on technology, prices, and the tax regime in force, more oil, and therefore more solution gas, can be available. According to some estimates, enhanced oil recovery might yield a further one-quarter of what can be produced by primary recovery

Table V-1

Natural Gas Reserves and Resources

(billion cubic metres at 1.1.85)

	Proved Reserves[1]	Additional Resources[2]	Current Gross[3] Annual Production
A. OECD			
North America	8 200	32 000- 42 000	
Of which: Canada	2 600	7 000-12 000	100
United States	5 600	25 000-30 000	550
OECD Europe	5 700	8 000- 16 000	
Of which: Netherlands	1 600	2 000	60
Norway	2 600	3 000-10 000	30
Others	1 500	3 000- 4 000	75
OECD Pacific	800	2 500- 3 500	
Of which: Australia	600	2 000- 2 500	13
Others	200	500- 1 000	5
Total OECD	14 700	42 500- 61 500	843
B. Centrally Planned Economies			
Of which: USSR	38 000	90 000-120 000	637
Others	1 300	1 000- 5 000	80
Total CPEs	39 300	91 000-125 000	717

Table V-1 (Continued)

Natural Gas Reserves and Resources
(billion cubic metres at 1.1.85)

	Proved Reserves[1]	Additional Resources[2]	Current Gross[3] Annual Production
C. Developing Countries			
Africa	5 500	+/-11 000	120(50)[4]
Of which: Algeria	3 100		
Others	2 400		
Asia	4 700	+/- 6 000	90(75)[4]
Of which: Indonesia	1 100		
Malaysia	1 400		
Others	2 200		
Latin America	5 300	+/- 8 000	120(75)[4]
Of which: Mexico	2 200		
Others	3 100		
Middle East	25 000 +	+/-44 000	110(50)[4]
of which: Iran	13 600		
Qatar	4 000 +		
Saudi Arabia	3 500 +		
Others	3 900		
Total Developing Countries	40 500 +	+/-69 000	440
Total World	94 500 +	200 000-260 000	2 000

1. **Sources:** OECD countries - government submissions.
   Sources: Other countries - International Gas Union, World Gas Conference (WGC, Munich), June 1985: Cedigaz; Oil and Gas Journal.
2. **Sources:** OECD countries - government submissions and IEA Secretariat estimates.
   Sources: Other countries - WGC, Munich.
3. Figures are approximate and gross, including flaring and re-injection.
4. Commercialised production shown in parentheses.

alone, and calculation of the ultimate resources of solution gas could be increased accordingly. Secondly, and more significantly, some types of sedimentary basins are more gas-prone than others — and these do not happen to be the same as the ones which are oil-prone. Much less exploratory work has been done in the gas-prone type of basins, simply because gas is less readily marketable than oil, being more difficult to store and transport. Delineation of gas resources world-wide is therefore, on balance, much less thoroughly developed than delineation of oil resources.

With these warnings on the limitations of the data in mind, Table V-1 gives some recent estimates of world proved reserves by region, as well as an assessment of total additional reserves and resources.

Active reserves in each of the OECD regions shown in the table will lie somewhere between the figures for "proved reserves" and the figures for "additional resources"[1]. It is from active reserves that future supplies are likely to flow. Many types of model have been developed to try to assess what a typical supply profile might be from any given reserves/resource base. Life-cycle, geologic-volumetric, and rate-of-effort models have perhaps been most widely used. No consensus exists as to which, if any, approach is the best way of relating reserves to actual supply in any given area. In the assessment of the supply of gas in the OECD regions in this study, no such modelling approach has been taken; instead the study relies on a descriptive assessment of the possibilities for supply through to 2010. Nevertheless, two fundamental conclusions may be drawn from reserves-supply modelling work which has been done in recent years. Both of these conclusions have important implications for policy making, given the geographical and political distribution of natural gas reserves and resources.

The first conclusion relates to the pattern of discovery of natural gas reserves. Typically, when a new gas province is discovered, it is the larger fields which tend to be discovered first, and the smaller ones later. There have been exceptions to this pattern but the experience of the petroleum industry over a hundred years has tended to confirm the rule.

---

1. Definitions are less certain for countries outside the OECD area, and some gas included in proved reserves may fall into the passive or sub-active category — notably in the Middle East and Africa, perhaps also in the Soviet Union.

Logically, therefore, since it is widely believed that a majority of the world's hydrocarbon-bearing sedimentary basins have already been or are in the course of being developed, it must be expected that the frequency of very large finds will tend to diminish. Based on existing known reserves, and on the probable development of supply in the main consuming regions, there is no shortage of natural gas available for the foreseeable future. In more mature gas provinces, however, maintaining an established level of supply will tend to require an increasing level of investment effort and cost.

The second conclusion, which gives more grounds for optimism, focusses on the dynamics of the development of gas reserves into gas supply. In past developments actual gas production from a given area of gas deposits has exceeded the rate which the quantity of known initial reserves alone indicated to be sustainable. Reference has already been made to the post-war experience of the United States in this regard. Improved seismic technology may now make estimation more accurate than it has been in the past, although the science remains inexact and areas of major uncertainty remain.

Figure V-2 shows for illustrative purposes the possible typical evolution of a gas province where reserves are assumed to be confirmed at a rate of only 50% of the estimated undiscovered resources, thus taking into account the tendency of large fields to be discovered first. Figure V-2 shows that if an initial level of production from the first discoveries in a new province is defined as 100, and the quantity of discovered gas in known fields also defined as 100, then over the first years of the province's life, both the rate of production and the volume of known gas will tend to increase. The actual rate of increase in supply will depend on a host of economic, technical, and political factors which determine whether a low or a high rate of annual production is desired.

In the light of the above review of the meaning of gas "reserves", and of the processes which take place in converting reserves into supply, it becomes clear that the size of proved reserves in any given country or region cannot be translated to mean that "gas reserves will last for x/y years", where x is a reserves figure and y is a current or estimated future annual production rate. A large proved reserves figure in country or region A, compared with a smaller figure in country or region B, does, however, indicate that gas reserves are more readily accessible in country A than in country B, probably at lower production costs. However, relative production costs are only part of the story when determining

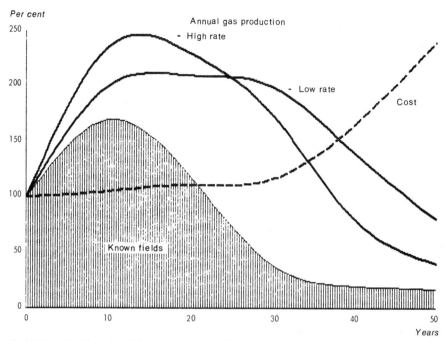

Figure V-2
Dynamics of Gas Province Development

Per cent

Annual gas production

250

- High rate

200

- Low rate

Cost

150

100

50

Known fields

0

0   10   20   30   40   50

Years

Coefficient of confirmation of forecast reserves = 0.5.

Source: A.A. Makarov, in Energy Reviews "Unified Gas Supply System of the USSR", OPA (Amsterdam), 1985.

whether or not gas reserves will be developed. Produced gas must also be transported to a consumer if it is to be developed for sale. Where this involves the construction of new infrastructure, such as gas gathering systems, transmission pipelines, and reticulation networks, the transport costs can be substantial, and may be very large relative to the raw production costs. Finally, if the delivery of gas from a low-cost producing region to consumers involves cross border trade, then political factors also come into play, and influence the rate at which reserves will be developed.

Figure V-3 shows the size of proved natural gas reserves in the main OECD regions as at the beginning of 1985. The cumulative production of each region since the early exploitation of gas in the last century is also shown. These quantities may be related conceptually to the relevant

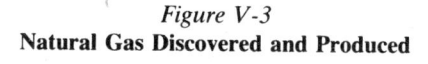

Figure V-3
Natural Gas Discovered and Produced

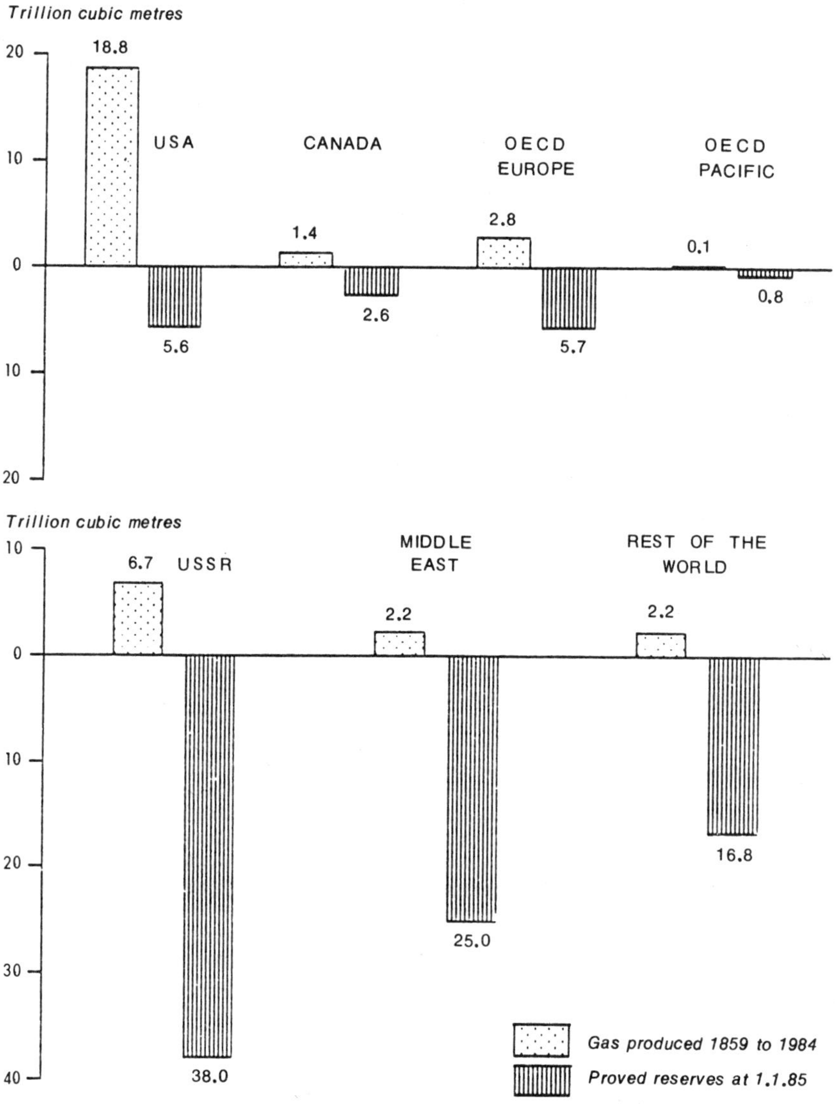

Trillion cubic metres
20
10
0
10
20
18.8
USA
5.6
CANADA
1.4
2.6
OECD
EUROPE
2.8
5.7
OECD
PACIFIC
0.1
0.8
Trillion cubic metres
10
0
10
20
30
40
6.7
USSR
38.0
MIDDLE
EAST
2.2
25.0
REST OF THE
WORLD
2.2
16.8
Gas produced 1859 to 1984
Proved reserves at 1.1.85

Source: Gas produced, BP Review of World Gas 1984; Proved reserves, Table V-1.

categories which were shown in Figure V-1. As explained above, future domestic supplies of gas in the OECD countries will come from a broader base of active reserves than just the proved reserves shown in Figure V-3. Figure V-3 also shows, for comparison, remaining "proved" reserves and quantities of produced gas in the Soviet Union and the Middle East. These figures indicate the availability of low production cost gas on a large scale in these two regions, at least some of which, following existing trading patterns, is likely to be consumed in OECD countries. One vital development on the world gas scene in the 1980s, which will affect supply prospects in the 21st Century, has been the completion of a transportation infrastructure, the unified gas supply system of the Soviet Union, which enables reserves in the Asian part of the Soviet Union to be delivered to Europe, including to OECD European countries. There is as yet no infrastructure to bring Middle Eastern gas to Europe, and only very limited LNG capacity (linking Abu Dhabi to Japan) between the Middle East and the OECD Pacific region. Transport costs are such a significant element in supply that without such infrastructure in place, even low cost Middle Eastern reserves must for the time being be considered "passive" as far as OECD markets are concerned.

B. The Costs of Gas Supply

Gas reserves have no value unless they can be delivered to someone who has a use for them. The costs of delivery can be divided into three parts:

- the costs of discovering and developing gas;
- the cost of transporting gas from where it is located to where it can be used;
- the cost of providing supply in a manner which the user is able to accept.

These may be termed respectively the production, transmission and storage/distribution costs of natural gas. The costs of taxation, as well as these technical costs, will be taken into account by those who develop and supply gas. It is important to note that actual supply options will be conditioned not only by costs. Other factors, such as prevailing pricing principles, competition between gas and other energy sources and the existing institutions and policies governing gas industries' operations could tend to lead to a ranking of supply options based on factors other than cost.

Storage and distribution costs are usually met by wholesale and retail distributors of gas in the OECD countries, except where producing gas fields themselves can act as storage by providing flexible supplies according to seasonal and daily needs. This section will exclude consideration of storage and distribution costs as well as fiscal costs and will consider trends in the technical aspect of the production and transmission costs of major existing and potential supplies of gas to OECD consumers and their implications for future supply.

Table V-2
1984 Gas Well Costs in United States Lower 48
(National Averages)

Well Depth in metres (feet)		Average Cost per Well (thousand 1984 dollars)	Depth Factor	Cost Factor (rounded)
380	(1 250)	44	1	1
760		96	2	2
1 520	(5 000)	207	4	5
2 290		460	6	10
3 050	(10 000)	939	8	20
3 810		1 679	10	40
4 570	(15 000)	2 824	12	65
5 330		5 000	14	110

Source: Oil and Gas Journal, December 31, 1984.

In North America, the supply of new gas in recent years has been assured principally by relatively small supplementary discoveries in mature gas provinces. Infill and extension drilling and drilling to new horizons in known fields and basins, rather than the development of large new mega-projects, have been the order of the day. The main element in production costs is the cost of new gas wells, and the trend in these costs gives the best clues to the evolution of future supply costs. 1985 well costs are shown in Table V-2. The capital cost of transmission has been less important because existing trunk transmission line infrastructure has been available to move new gas supplies in established producing areas to traditional markets. Total gas sales in the important United States market have declined since 1973, so that there has been relatively little investment in major transmission projects. In western

Canada, however, the construction of the so-called "Prebuild"[1] pipeline has been an exception to the general North American rule, and has been a significant element in total gas supply costs.

Onshore gas well costs in the lower 48 States of the United States show clearly the correlation between costs and the depth of wells drilled. At well depths of more than 1 500 metres (5 000 feet), costs increase significantly more than proportionally to increasing depth. At 3 000 metres, average costs are more than twelve times the average cost of 1 500 metre wells. In those parts of the United States where drilling has gone below 5 000 metres (the Rocky Mountains, mid Continent and southeastern states) typical costs can soar to more than 20 times the cost of shallow wells. Table V-2 illustrates the exponential relationship of well depth to cost, based on averages for the lower 48 States.

Future trends in United States gas supply costs are likely to be strongly upwards if it is necessary to drill deeper and deeper to find substantial new onshore supplies. According to latest estimates, only some 60% of drilling by producers in 1985 is likely to be to depths of less than 1 500 metres (5 000 feet) onshore[2]. Other drilling will either be offshore in the waters of the continental shelf, or to depths greater than 1 500 metres, or in tight sand formations and Devonian shales, where production of gas is substantially more expensive. The trend towards more expensive and difficult drilling is likely to continue, especially as long as price controls remain on gas classified under sections 104 and 106 of the Natural Gas Policy Act. These controls frequently have had the effect of making it more economic for producers to drill deeper, more expensive wells to obtain new gas supplies rather than to drill to known deposits in shallower horizons. There is correspondingly a substantial amount of gas in the lower 48 which is shut in "behind the pipe" by the remaining price controls. Some estimates put the volume of this gas at between 750 and 1 400 bcm [27 to 49 trillion cubic feet (TCF)[3]].

---

1. The name derives from the original intention to make this pipeline a part of an Alaska-to-United States lower 48 pipeline, this section being "pre-built" to deliver smaller quantities of Canadian gas.

2. Source: Energy Information Administration, Natural Gas monthly, May 1985. Table 10, "Well Determination Filings by Category".

3. Source: United States Department of Energy, "Increasing Competition in the Natural Gas Market, The Second Report Required by Section 123 of the NGPA of 1978", DOE/PE-0069 (1985), pages 129-143.

If remaining price controls were to be removed in the medium term, geological factors are still likely to drive the trend in production costs upwards in the United States in the longer term. Future gas supplies will come increasingly from offshore developments, especially from deeper waters in the Outer Continental Shelf. The timing of this move to deeper waters will be determined essentially by two factors on the supply side — the amount of Canadian gas available for export to the United States in the medium-term, and the possibilities of legislative change releasing the gas "behind-the-pipe" in the United States itself.

In the transmission sector, there have been no major new capital investments in recent years. Some new market areas may require new trunk pipelines in the near- to medium-term future, however. In particular, enhanced heavy oil recovery projects in California, which have substantial steam-raising needs and tight environmental controls, are likely to require major new supplies of gas from the southwest central states or from Canada or the Rocky Mountain overthrust belt. Further penetration of the northeastern states' space heating market is also likely to require new infrastructure investment to move Canadian gas into the United States — either directly supplying the northeast, or by exchanging Canadian gas into the Midwest against re-routed Texas and Louisiana supplies. Proposals have been tabled for each of these projects. In the longer-term, transmission costs could become a more important component of general United States gas supply costs if either some form of the Alaskan Natural Gas Transmission System (ANGTS) were to be revived or if new LNG import projects were needed to cover a significant part of United States demand.

The costs of ANGTS have been estimated to exceed $40 billion (in 1984 dollars). It is difficult to determine the volume of gas from alternative sources (LNG imports from North and West Africa, Latin America and the Pacific or unconventional sources in the lower 48) which could also be made available at this cost to United States consumers. Based on current trends in supply availability and depending on Canadian export policy, at least one of these more expensive options (ANGTS, LNG or unconventional sources) or some combination of all three, appears likely to be required for United States gas supply by the first decade of the next century.

In European OECD countries, present and future trunk transmission investments, as well as production costs, are critical elements in the cost of gas supply from various sources. In the last five years, new infrastructure has been needed to bring to market large gas deposits in Norway, the United Kingdom, the Soviet Union and Algeria. Much of this has included expensive offshore work. Estimates of the capital costs of these projects are shown in Table V-3.

Table V-3
Major Gas Transmission Projects for Europe, 1980-85

Pipeline	Date	Supplying Fields	Estimated Approximate Cost (US$ billion)
Transmediterranean	1979-83	Hassi R'Mel	1.6[1]
FLAGS	1980-83	Brent and others	0.6
DONG	1979-83	4 Danish fields	0.8
Statpipe	1981-85	Statfjord-Heimdal-Gulfaks	3.5
Brotherhood/ Northern Lights	1980-86	Urengoi and other Tyumen fields	16.0[2]
MEGAL	1978-86	Various Soviet Union	1.1

Sources: Various (World Bank, Arab Oil and Gas, company annual reports and others).
1. Costs incurred in Algeria, Tunisia and Sicily Channel.
2. Does not reflect full cost of these two pipeline systems, which form part of the Integrated Natural Gas Supply System (INGSS) of the Soviet Union. Estimate relates to export-linked pipelines for East and West Europe.

It should be stressed that these estimates are for the transmission systems alone; they do not include the costs of developing gas production from the related fields. These investments have followed similar major projects in the 1970s — the Norpipe line from Ekofisk to Emden and the Frigg lines to St. Fergus in the North Sea, the Orenburg to Uzhgorod line in the Soviet Union and Algeria's LNG projects at Skikda and Arzew. New gas supply projects to Europe in the 1970s permitted rapid growth in gas sales and consumption, as they effectively supplemented growing supplies from the Netherlands as new markets were being penetrated. The 1980s investments, however, have been made during a period in which there has been little or no net growth in gas demand: OECD European countries' consumption of 212 bcm of gas in 1984 compares with 209 bcm in 1979, which contrasts with 140 bcm in 1973. Some investments (such as the Transmediterranean pipeline) have been associated with the expansion of national or regional markets, but most investments in this period have been designed to offset expectations of declining supplies from traditional sources — notably the Netherlands and the Southern Basin of the United Kingdom North Sea, but also French, German and Italian domestic onshore production.

An economic assessment of the cost of gas supply to OECD European countries must allow for the element of the capital cost of new transmission systems in the case of new supplies. For some of these supplies, especially the offshore ones, substantial production investment costs must also be included. It is difficult to assess these costs, because of the interdependence of oil and gas developments (e.g. in the case of the Statfjord and Brent fields and other oil fields which supply or will supply large volumes of gas to the Statpipe and Far North Liquids and Associated Gas (FLAGS) systems respectively) and the arbitrary nature of any allocation of costs between them. However, in the case of some of the major potential future new supplies of gas to Europe, where new fields may be developed, the main interest may lie with non-associated gas, gas/condensate reservoirs, or fields where the gas reserves are much larger than the developable oil reserves. In particular this would be the case with the Troll and other fields offshore Norway, gas condensate reservoirs offshore the United Kingdom, Nigeria's Rivers' State reserves, and the Qatari North Dome field. In all these cases, the economics of development will essentially depend on the relationship between the production cost and the realisable price for gas with only some offsetting allowances for associated liquids development.

The costs of existing gas supply and of incremental supplies delivered from existing production capacity through transmission infrastructure already in place, can more appropriately be judged either with respect to marginal production and transmission costs or with respect to some depletion-related or reserves-replacement opportunity cost.

For geographical, geological and technical reasons, different regions face widely differing levels of cost in developing their gas resources. Tables V-4 and V-5 set out a systematic ranking of the various major cost elements for each of the countries who are suppliers or potential suppliers of gas to Western Europe over the next 25 years. Effects of taxation are excluded from these tables. Table V-4 is based on the assumption that all suppliers can cover fully built up costs, including capital charges. Table V-5 shows only variable cost recovery, such as might be induced by excess supply capacity, or gas-on-gas competition. In each case, there is an implied ranking of suppliers in terms of economic attractiveness baed on costs alone. By assigning a numerical value to the cost factors in each of the categories, this ranking has been formalized in the tables. A comparison of the two tables can give insights into the relative cost positions of various potential suppliers of incremental gas — some of which could make gas available on a variable cost basis, where others must face fully built up costs.

Table V-4
Indicative Ranking of European Natural Gas Supply Costs
Basis: Fully Built Up Cost

	Production Costs	Gathering Costs	Processing Costs	Transportation Costs	"Score"[2]	Rank
Algeria	low	low/medium	high (LNG) low (pipe)	high (LNG) medium (pipe)	5- 9	3=
Austria/Germany/ Italy	low/medium	low	low/medium	low/medium	4- 6	2=
Netherlands	low	low/medium	low	low	4- 5	1
Nigeria (LNG)[1]	low	medium	high	high	9	5=
Norway - North Sea	high	low	low	medium	7	3=
- Troms[1]	high	low	high (LNG) low (pipe)	high	8-10	5=
Qatar (LNG)[1]	low	low	high	very high	9	5=
Soviet Union	low	low/medium	low	high/very high	6- 8	3=
United Kingdom - Southern Basin	low/medium	low/medium	low	low	4- 6	2=
- N. & C.N.Sea	medium/high	low/medium	low	medium	6- 9	4

1. Projects in planning stage. No gas presently supplied from these sources.
2. "Score" calculated by assigning a value of one to each "low", two to "medium", three to "high", four to "very high" in each cost category.

Source: IEA Secretariat.

Table V-5
Indicative Ranking of European Natural Gas Supply Costs
Basis: Variable Cost

	Production Costs	Gathering Costs	Processing Costs	Transportation Costs	"Score"[2]	Rank
Algeria	low	low	low/medium	low/medium	4-6	3=
Austria/Germany						
Italy	low	low	low/medium	low	4-5	2=
Netherlands	low	low	low	low	4	1=
Nigeria (LNG)[1]	-	-	-	-	-	-
Norway						
- North Sea	low/medium	low	low	low/medium	4-7	3=
- Troms[1]	-	-	-	-	-	-
Qatar (LNG)[1]	-	-	-	-	-	-
Soviet Union	low	low	low	low/medium	4-5	2=
United Kingdom						
- Soutern Basin	low	low	low	low	4	1=
- N. & C.N.Sea	low/medium	low	low	low	4-5	2=

1. Projects in planning stage. No gas presently supplied from these sources.
2. "Score" calculated by assigning a value of one to each "low", two to "medium", three to "high" and four to "very high" in each cost category.

Source: IEA Secretariat.

The cost of gas supply to Japan is heavily conditioned by the high fixed costs of the various LNG "chains" which have been constructed to serve the market (in Indonesia, Malaysia, Brunei, Abu Dhabi and Alaska). Project costs for a 6 million metric ton per year liquefaction plant, plus production, gas-gathering and treatment facilities and port facilities and ships to deliver the LNG, can typically be about $7 to $8 billion. Some of these projects are still at an early stage of repaying the loans with which they were financed; one new project is under construction (in Australia) and others are planned, which will increase the volume of gas supplied to Japan as the market grows. For the medium term, fully built up cost recovery will continue to be the relevant criterion in calculating supply costs for this market. Longer term, if it is geologically, technically and commercially feasible to extend the life of supply contracts after the equipment in an LNG chain has been fully depreciated, the question may arise of the availability of gas at lower cost.

C. Supply Options for OECD Europe

(i) Expected Supply

Indigenous production of natural gas for use in the country of production currently accounts for just over half of the OECD European countries' total supply at 110 bcm out of 212 bcm (1984). This is likely to be sustained through to 1990 at a level of about 100 to 120 bcm, based on extrapolations from existing producing and contracted fields. Actual production in any given year will depend on the supply and demand match obtained by balancing flexibility in indigenous supplies with other contracts. The Netherlands and the United Kingdom, with 35 and 37 bcm respectively, are the largest producers of gas for their own national markets, followed by the Federal Republic of Germany and Italy with 15 and 12 bcm respectively. These four producers account for almost 90% of production for own national use in Western Europe, with Austria, Denmark, France, Ireland and Spain making up the balance.

By the year 2000, based on current knowledge and existing producer-transmission company contracts, these supplies could be in the range of 75 to 115 bcm. The range of uncertainty applies particularly to the United Kingdom, where a great deal will depend on the ability of the British Gas Corporation or other United Kingdom purchasers to contract for new supplies from the United Kingdom Continental Shelf.

Annual volumes of about 15 bcm are under negotiation between BGC and North Sea producers and further volumes are expected to be on offer in due course. The current Dutch policy ensures that the Netherlands will remain capable of fully supplying its domestic demand through to 2010 and beyond, supplemented as now with small volumes of Norwegian gas. If the United Kingdom Continental Shelf is to continue to provide the bulk of United Kingdom supplies to 2010, however, smaller, deeper, and more complex gas deposits in the North Sea will have to become more economic to develop than is currently the case. This may be possible as the price, technological and fiscal environments change over time.

The Netherlands and Norway are the main sources of additional contracted supplies in intra-OECD European trade. Denmark will also supply small quantities of gas under contracts to Sweden and the Federal Republic of Germany. Contracts made in 1985 between the Netherlands and buyers in Belgium, France, Italy and the Federal Republic of Germany have made additional volumes of Dutch gas available — in some cases potentially as far into the future as the year 2010. The total volume of Dutch gas committed to export at 1st January 1985 was about 550 bcm[1], which effectively guarantees a continued Dutch presence in the market through the next 25 years. Norway's presence as a supplier of gas to Europe is based on the sale of large gas deposits from two areas, Frigg and Ekofisk, supplemented from 1985 by a third major sale of gas from the Statfjord field through the Statpipe system. Gas from two other fields, Heimdal and Gullfaks, starts flowing through Statpipe in 1986 and will continue through the 1990s according to existing contracts, but there will nevertheless be a steady decline in the total supply of Norwegian natural gas in the 1990s unless new contracts are concluded. The existing Norwegian commitment is for some 290 bcm of gas, and annual volumes of about a further 15 bcm from Troll and other Norwegian fields are under discussion with continental European buyers. Without new Norwegian supplies, intra-OECD trade in gas could decline from around 50 to 59 bcm in 1990 to around 24 to 32 bcm in 2000. New Norwegian supplies to continental Europe could boost the 2000 figure to around 45 to 50 bcm, and deliveries could be sustained at that level or even higher given the present reserve base.

---

1. Defined according to the standard calorific content of gas used in this study, at 9 500 kilocalories per cubic metre. This corresponds to 625 bcm of Groningen-quality gas.

Contracted imports from other sources, Algeria, Libya and the Soviet Union, make up the balance between projected supply and demand through to the mid-1990s. Algeria has 550 bcm of gas committed to European buyers; the precise timing and volume of deliveries in the early 1990s will depend on the ability of the buyers to implement existing contracts as final markets develop. The recently resolved dispute with Spain has extended Algerian supplies to that country through to 2004. The Soviet Union has between 850 and 1 000 bcm of gas available under existing contracts for OECD European buyers, depending on the flexibility of offtake and supply which is opted for. Supplies from the Soviet Union will increase from the level of about 25 bcm established in the early 1980s to between 40 and 48 bcm by 1990 and to 43 to 55 bcm by 2000 as gas contracted under the Soyuzgazeksport-IV and other 1980s contracts builds up to full volumes. In addition to supplying gas to established trading partners, the Soviet Union will begin to send gas to Turkey (up to 6 bcm per annum), and possibly Greece, based on initial undertakings made in 1983 and 1984. Discussions are underway between Turkey and certain Middle Eastern countries about other potential supplies. Greece and Algeria have also explored the possibility of LNG deliveries.

The sum of all already identified supplies — including firmly contracted imports and gas which is expected with reasonable confidence to be made available from active indigenous reserves — is shown in Table V-6a. Actual supply is likely to be made up of a mixture of the high end of the ranges shown for some of the sources and the low end of the ranges for others, depending on operational conditions at any given time and will of course be supplemented by other supplies to be contracted for in due time, as described in section (ii) below. For the range of gas demand considered in this study, the extreme lower and upper aggregate ranges of supply therefore have a low probability attached to them. They represent the operational and contractual room for manoeuvre in the event of unanticipated changes in the general energy environment.

Figure V-4 superimposes the demand range forecast for benchmark years to 2010 on this range of firmly contracted and likely indigenous supplies. New contracts will be required in the near future that will allow for the delivery of new gas supplies from the mid-1990s. The actual timing of deliveries will depend on the degree to which supplies not lifted in the early 1990s, because of demand constraints, will be rolled forward into subsequent years.

Table V-6a
Natural Gas Supplies for OECD Europe
(billion cubic metres)

	1973	1985[1]	1990	2000	2010
Indigenous Production for Local Use					
Austria	2.1	1.2	1	0.5	-
Denmark	-	1.0	2	2	1- 2
France	7.4	5.2	3	-	-
Germany	17.7	16.4	14- 16	10- 15	6- 12
Ireland	-	2.2	2	2	1- 2
Italy	14.8	12.8	10- 12	10- 12	8- 10
Netherlands	33.3	33.5	31- 35	31- 35	31- 35
Spain	-	0.3	2	1.5	-
United Kingdom	28.6	40.5	35- 41	35- 55	10- 35
Sub-Total	103.9	113.1	100-114	92-123	57- 96
Contracted intra-OECD Trade					
Denmark	-	0.5	1	1	-
Netherlands	29.6	35.0	25- 30	20- 25	0- 10
Norway	-	24.0	24- 28	3- 6	-
Sub-Total	29.6	59.5	50- 59	24- 32	0- 10
Contracted non-OECD Imports					
Algeria[2]	2.2	19.0	23- 29	24- 30	-
Libya	2.4	0.4	1- 2	-	-
Soviet Union	2.2	26.6	39- 48	43- 55	-
Sub-Total	6.8	46.0	63- 79	67- 85	-
Possible Renewals and Extensions					
Denmark	-	-	-	1- 3	1- 3
Norway	-	-	-	18- 21	25- 45
Algeria	-	-	-	-	24- 30
Soviet Union	-	-	-	-	43- 55
Sub-Total				19- 24	93-133

1. Provisional and partly estimated.
2. Pending outcome of 1986 renegotiations with consumers in Belgium, France and Italy.

Source: IEA Secretariat.

Table V-6b
Natural Gas Supplies for OECD Europe
(billion cubic metres)

	1973	1985[1]	1990	2000	2010
Sub-Totals					
Indigenous production for local use	103.9	113.1	100-114	92-123	57- 96
Contracted intra-OECD trade	29.6	59.5	50- 59	24- 32	0- 10
Contracted non-OECD imports	6.8	46.0	63- 79	67- 85	-
Possible renewals and extensions	-	-	-	19- 24	93-133
TOTAL	140.3	218.6	213-252	202-264	150-239

1. Provisional and partly estimated.

Source: IEA Secretariat.

(ii) *Possible Sources of Additional Supply*

Additional supplies may come from a number of sources. Indigenous production for local use can be promoted by ensuring that the fiscal climate, exploration terms and gas sale and purchase conditions are attractive to producers. Already in the Netherlands, the contractual position of Gasunie, which undertakes to purchase at market prices any gas developed and offered onshore or offshore the Netherlands, acts as a stimulus to the development even of relatively small accumulations of gas (down to 1 or 2 bcm). Conditions in the other major offshore producers are evolving over time, as fiscal regimes adapt to the prevailing energy price and risk environment.

The Netherlands, however, exercises what is effectively a depletion policy, under which export commitments are not allowed to exceed a volume of reserves in excess of that volume which can conservatively be considered available to cover at least 25 years of future domestic gas demand. These volumes are re-assessed on an annual basis. Although

Dutch gas is proven available, further supplies in addition to those contracted for export in 1985, would require re-assessment at a later date by the Dutch Government and its partners of the desirability of selling additional volumes of gas from known reserves. Future discoveries, extensions, or increases in confidence levels about the quantity of existing reserves could in principle be sources of additional gas without requiring changes in policy or diminishing the security reserve in the ground. In the light of recently agreed extensions to existing export contracts, consideration of further developments is clearly premature.

In Norway, exploratory activities on the Continental Shelf have so far proven gas reserves of more than 2 500 bcm. This provides a potential reserve base for gas supplies to Western Europe far into the next century. The Troll field alone has reserves of around 1 200 to 1 300 bcm, which would be sufficient to supply 10 to 15% of projected West European demand for forty years. Progress continues to be made on the definition of development options for Troll. Talks with potential buyers on the European continent concerning deliveries of gas from Troll and other fields are proceeding. Development of the Troll field would entail high financial exposure and risk for participants, although a development time of seven, rather than nine, years is now believed to be possible because of recent technological advances. If Norway were to seek to maintain sales of 25 to 30 bcm of gas to Europe in the year 2000, then at the lower end of the projected difference between supply and demand, new Norwegian developments alone could be sufficient to match the two. Early decisions on new developments are, however, necessary if sales are to be maintained.

Algeria could make available additional volumes of gas over and above existing contracts. Extra compression and looping key sections of the Transmediterranean pipeline could technically permit gas to be moved north to supply additional demand in Italy and other countries. Spare LNG capacity exists, as a result of the cancellation of the El Paso contract to the United States in 1980 and the suspension of liftings by Trunkline at the end of 1983 and by Distrigas of Boston in 1985, as well as from a lower than expected build up in deliveries to some European buyers. This capacity probably does not exceed 6 bcm, taking into account both the reduced operating ability of the Skikda liquefaction plant and difficulties with operations of Arzew. Major overhauls could be needed on some other aging equipment if extra gas were to be offered as LNG to Europe in the 1990s. The particular indexation provisions of the Algerian contracts — linking gas to crude oil prices — imply that price developments are not necessarily in line with market developments. Furthermore, the absolute level of the Algerian gas price has limited the rate of penetration of Algerian gas on the European markets.

Figure V-4 also shows the profile of the supplies to 2010 which would result from an extension of current contracts from Algeria, and the Soviet Union, and from possible Norwegian supplies. As to the Soviet Union, in the 15 years between 1995 and 2010, reserves are not likely to be seen as a constraint when considering exports to Western Europe. There is likely to continue to be excess delivery capacity to the western border of the Soviet Union, where the gas export pipelines were sized to accommodate the now defunct tripartite proposal under the IGAT-II deal to exchange Iranian gas for Western Europe with additional Soviet

Figure V-4
Europe
Natural Gas Demand and Supply 1970-2010

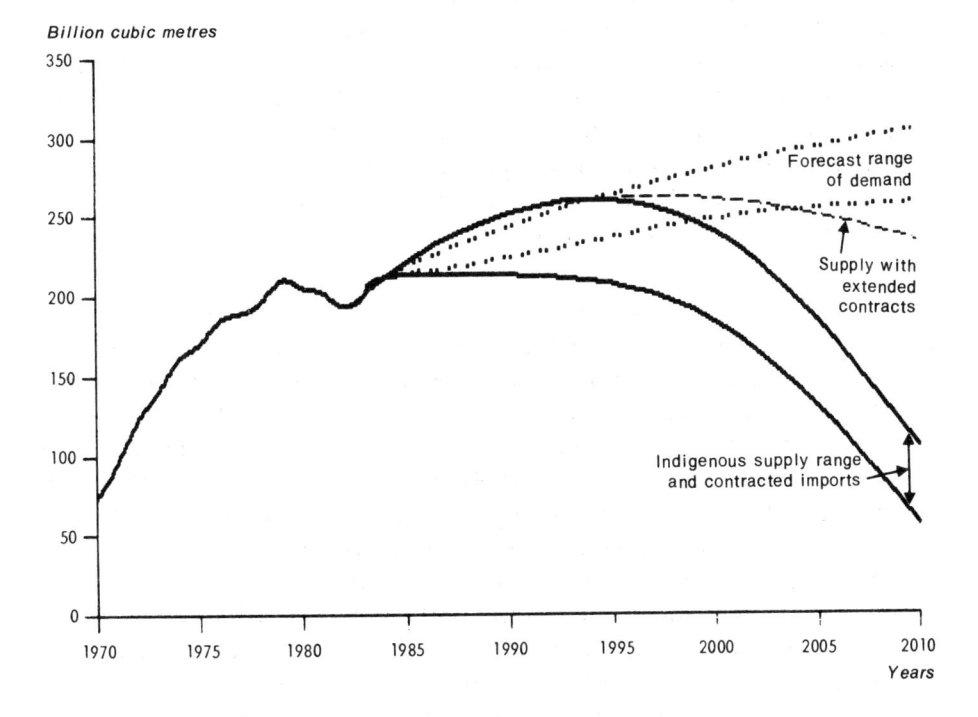

Source: IEA Secretariat.

deliveries. However, the transportation capacity through Czechoslovakia and through the Trans Austria Gasleitung (TAG) in Austria to the MEGAL system in Germany will when completed only accommodate a reduced volume of around 50 bcm per annum.

Other sources of gas for Europe could include new discoveries in prospective areas such as the Celtic Sea off Ireland, deep water areas north and west of the United Kingdom, the Italian Adriatic and onshore basins in Germany and elsewhere. Large volumes could also come from further afield than the existing four major suppliers described above. LNG projects have been proposed and are still under review in Nigeria and Qatar. One option for developing Norwegian reserves could be a liquefaction plant in the north of the country. There could be room for any one of these projects in the European gas supplies if they can be developed to supply gas competitively with other fuels. Proposals to develop a pilot project to bring LNG from Arctic Canada to Europe have been shelved and there is some doubt about the adequacy of the reserve base in the Cameroons. Other long-haul LNG projects are unlikely to advance beyond the drawing board stage unless and until the economic environment for energy supply projects changes considerably. The economics of pipeline supply projects over distances greater than 3 000 to 6 000 kilometres (depending on the assumptions) are less attractive even than the economics of equivalent LNG projects[1].

(iii) Pricing Principles and Gas Supply

Gas supply is, and will continue to be, strongly influenced by the principles which govern prices paid to producers. In general terms, three distinct sets of principles have been either applied or proposed at various times by various parties to the pricing of wholesale gas supplies to West European transmission companies. These principles may be summarised as suggesting that gas should be priced either:

(i) so as to be at or near parity with crude oil export prices, or

(ii) so as to be competitive with the final consumer's alternative non-gas fuels, or

(iii) so as to reflect the historic costs of gas production.

---

1. See "Natural Gas Prospects to 2000", OECD/IEA 1982, page 125 citing G. Bonfiglioli and F. Cima, "Economics of Gas Utilisation in Different Fields", 1980.

The way in which the different segments of the gas industry are organised, including the relative strength of buyers (transmission companies) and suppliers (producing companies), will influence the type of pricing principle which is likely to be adopted. This will also be influenced by the degree of mutual interest perceived by buyers and sellers in the stability of long-term relationships between them.

If the third of the above principles — historic production costs — is applied, then there is a risk that indigenous gas supplies will be underdeveloped over the long term. Prices based on covering historic costs may be inadequate to stimulate development of marginal new fields. The result can be that gas resources in the area to which this pricing principle is applied are left unused, even undiscovered, because there may be no incentive to explore for or develop new gas. As an example, development of United Kingdom Continental Shelf gas in the 1970s may have suffered from this phenomenon; changed policies make it more likely that this gas will be developed in the late 1980s and 1990s.

If the first principle — crude oil parity — is applied, then although supply may be assured, there is a risk that gas loses market share against other fuels, especially against oil, whose lower transport, storage and distribution costs would give it a competitive advantage over gas at the point of consumption. This will be the case wherever the price of consumers' alternative fuels is lower than the price of delivered gas. In other parts of the world, where crude oil itself may have been a competing fuel in end use (for example, in Japanese power stations) the principle of crude oil equivalent pricing, on a CIF basis, has been capable of yielding a price for imported gas which is competitive with consumers' alternative fuels and which is therefore consistent with gas maintaining its market share. This has not been the case in any European country.

The second principle — that the price of gas should be generally competitive with the prices of non-gas fuels — has been the most widely applied in Western Europe. In the Netherlands, for example, there is a policy that the transmission company will assure purchase of all gas developed at passed-through market prices. This is combined with a policy of keeping in reserve the low-historic cost Groningen field for domestic security and for seasonal flexibility. As a result, new field developments in the Netherlands can be assured, whilst at the same time the level of prices permits the development of new gas resources from other suppliers to be stimulated. This will continue to be the case as long as these policies are in place, and as long as those other suppliers' resources can be delivered competitively with the final consumers' alternative fuels. Elsewhere in continental Europe, transmission companies which are willing to pass through to their suppliers prices based on this principle have stimulated both domestic production and widening diversity of supply. As long as alternative fuel prices exceed the cost of producing, supplying and distributing new gas, this principle should ensure that adequate gas supply is available to meet stable or rising demand. It may also leave an element of economic rent where some suppliers have lower overall costs of supply than the marginal gas development.

The West European gas industry is presently organised in such a way as to make application of this pricing principle sustainable. An important aspect of government policy, in overseeing the industry environment, will be to ensure that this remains so and that the risks of narrowing the supply base by excessive reliance on lower than replacement cost sources of supply can be avoided. For it is clear from Table V-5 that, on a variable cost basis, the main suppliers of gas to Europe may face very similar cost conditions. In the context of excess supply capacity, this could lead to intense gas-on-gas competition, and to prices moving below consumers' alternative fuel costs, towards those based on historic costs of development or on variable costs of production. Such prices would not necessarily reflect the true market value of gas, measured in terms of assuring consumers of continuing supply. Gas contracts are made long-term on stable principles in order to resolve this problem. Equally, it is clear from Table V-4, that prices which give fully built-up cost recovery to ensure development of gas from new suppliers (such as deep water, northern Norway, West Africa or the Middle East) might leave room for substantial economic rents for other, lower-cost producers. If buyers are to be willing to pay such prices, instead of prices based on production costs from the spare supply capacity of lower cost producers, then the industry must continue to be organised in such a way as to make its long-term relationships stable and alternative fuel pricing concepts acceptable.

In terms of matching potential supplies to potential demand, the critical issues may be summarised as:

(i) What level of price, at the input to national transmission grids, can importers afford to pay and still be able to sell the contracted quantities of natural gas in a competitive market?

(ii) Do these prices, after deducting production and transportation costs, result in a return to producers adequate to encourage exploration for and production of gas for export?

As a first step, these questions can be approached by calculating netback border values for gas. These can then be compared with estimated costs for possible incremental gas supply projects. An illustration of such an approach, using average competitive fuel values and estimates of distribution costs in the various sectors of consumption, is given in Appendix III. The calculation is highly generalised, to be representative of Europe as a whole, and is applicable to no particular set of national or market conditions. It excludes any reference to fiscal terms. Furthermore, in view of the industry's risk and reward considerations, a positive result would be a necessary, but not a sufficient, condition for deciding whether a project can go ahead. The conclusions of such an analysis will vary directly, of course, with the cost, fuel price and demand assumptions. In the event, these variables will almost certainly differ from the assumptions.

The approach taken here is appropriate for long-run analysis, but cannot necessarily be applied to individual trade proposals. The issues involved in any particular case are complex and in general, the amount that importers can reasonably be expected to pay will be affected by many considerations, which have not been quantitatively factored into this study's calculations, such as the following:

- *the age, capacity and expected utilisation rate of existing distribution systems.* Where new systems are required, the amortisation of the additional investment cost will lower the price that can be paid for new sources of supply.

- *the relative share of different market sectors, with differing seasonal demands, in the market for existing and other incremental natural gas supplies.* The competitive price at the burner-tip is higher for residential than for most industrial use. However, the costs of supplying residential demand are substantially higher, both because of the higher fixed costs of distribution and because of the need for seasonal and daily storage to meet variations in demand. In the case of interruptible industrial supply contracts, where the price to the final user must be low, there may be an offsetting "benefit" (rather than a cost) to the distributor because the ability to interrupt supply forestalls the need for incremental storage to meet seasonally variable demand. The distribution and storage cost effects will tend to bring closer together the value of so-called "high value" residential markets and so-called "low value" industrial markets.

- *the structure of domestic taxation on energy products.* Where higher consumption taxes are levied on competing fuels than on gas, individual utilities may be able to offer a higher price for gas than would be the case if both fuels were taxed equally. From a national point of view, however, this would result in transferring the value of these tax revenues out of the importing country. If there are production taxes on gas in the producer country, the revenues are effectively transferred to that country.

- *the market position of natural gas in the overall energy economy.* If gas is to increase its market share faster than the natural rate of replacement of burner equipment, then a price which is competitive with an alternative fuel must be defined to include an extra differential to amortize early conversion of capital equipment (burners and boilers) to gas from that other fuel. In the long-run analysis undertaken in this study, natural gas replacement needs are assumed to determine the rate of penetration. Local conditions, especially in new market areas, may differ radically at any given time. This general analysis should not therefore be interpreted as applicable to any particular market at a particular time.

In addition to considering prices which can be realised, producers will base decisions on whether to export gas on the following factors:

- *Costs of production:* gas production in some countries is moving to greater target depths, greater offshore water depths, more difficult climatic conditions, and more complicated and smaller geological structures. Financial and management risks increase with project size and must be traded off against the scale economies which are usually associated with gas supply projects.

- *Alternative uses:* gas may be sold in the producing country as a fuel or feedstock for developing local industry to manufacture higher value-added products. It may also be used simply as a low cost fuel for local electricity generation. Such uses may yield better wellhead netbacks than, for example, a highly capital-intensive, high foreign exchange component LNG or long-distance pipeline export project. Such alternative domestic uses may be particularly valuable where the gas displaces imported fuels or frees domestically produced, and cheaper to transport, oil for export.

— *Fiscal regime:* fiscal regimes in both producer and consumer countries can make the difference between a technically viable export project and a commercially viable one.

Two important conclusions emerge from the analysis of Appendix III, irrespective of the actual values used and forecasting price assumptions. First, the netback border value of gas will fall relatively to the price of oil as oil prices fall and rise relative to the price of oil as oil prices rise — even where consumer price relationships follow the profile used in this study. Figure V-5 illustrates this relationship, using the values calculated

Figure V-5
**Netback Border Values for Natural Gas
in Europe 1984-2010[1]**

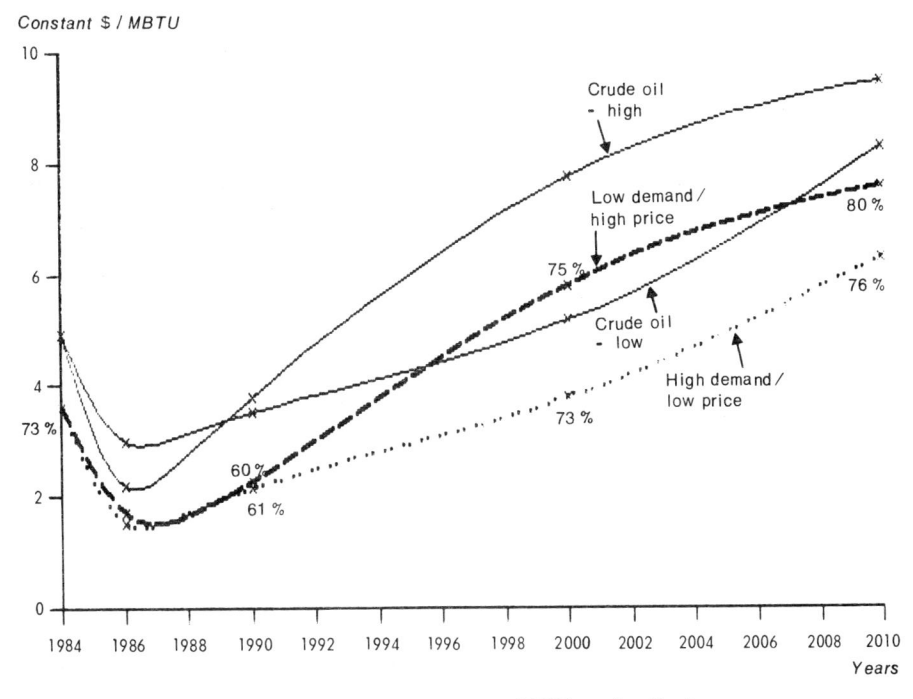

1. Percentages are border values of gas as per cent of FOB crude oil prices.

Source: IEA Secretariat.

in Appendix III. This is because the element of distribution costs is essentially fixed, and these account for a larger share of the total cost of natural gas than they do of crude oil and oil products. It is especially true when incremental storage is required to meet a high share of seasonally variable demand (mainly in the residential and commercial sector), or to ensure security of supply for a larger total volume of demand in all sectors. Under these circumstances, total distribution costs will increase even where transmission and reticulation networks have been installed and fully amortized.

Second, where residential and commercial sector gas demand growth increases relatively faster than average demand growth, the corresponding increase in the competitive fuel equivalent value will always be offset to some extent by an increase in average distribution costs. Although the addition of marginal residential customers (the so-called "premium" market) may increase average short-term realisations where a mature, but not fully-utilised reticulation network exists, over the long-term the average value of this market must be reduced by the requirement to balance load. Netback border values will reflect this phenomenon so that even in the case where industrial demand is relatively stagnant (the low growth, high oil price case) there are effectively dimishing returns in terms of value to incremental demand growth from other sectors.

D.   Supply Options for the United States

(i)   Introduction

Future supplies of gas for the United States will come from a number of sources. The major possible sources are usually classified as:

- conventional lower 48 States;
- incremental tight formation in the lower 48 States;
- Canada;
- Mexico.

These could be supplemented by supplies from:

- Alaska;
- imported LNG from diverse sources;

-   coal gasification;

-   synthetic natural gas (SNG) from naphtha, condensates or heavy residues;

-   non-conventional sources (coal seam gas, biomass and waste conversion, geopressured gas, shale oil gasification).

Supply options for the United States are dominated by two main factors — the legislative/regulatory environment and the geological maturity of many of the main producing provinces in the lower 48 States. The legislative environment has evolved towards the decontrol of wellhead prices since the enactment of the Natural Gas Policy Act in 1978. Wellhead price controls remain for gas defined in certain categories of this Act, and will remain in place until gas under contract in these categories ceases to flow on contract expiry. Other gas, which would have to be classified in these categories if it were produced, but which is not produced because it is uneconomic under the present price ceilings, will remain in the the ground and may never be available as supply unless there is some legislative or regulatory change to release it.

(ii) Pricing Principles

Figure V-6 illustrates the possible effect on supply of complete deregulation of wellhead prices of this category of gas (usually referred to as "old" gas). Included in this category is gas classified under sections 104, 105 and 106 of the 1978 Natural Gas Policy Act. Figures V-6a and V-6b show the composition of major interstate pipelines' gas supplies in 1981 and 1984, as filed with the United States regulatory authorities divided between old and other, "new", or decontrolled gas. These may be considered typical of the interstate gas market as a whole. In response to market developments and the economic effects of wellhead price constraints, the volume of old gas supplied to pipelines has declined since 1981. At the same time, the price of the highest cost gas (gas classified under Section 107 of the Natural Gas Policy Act (NGPA)) has come down, as the scope for rolling-in unusually high cost gas has been reduced with the falling share of old gas in the total. The volume of high cost gas has increased, however.

Deregulation has proceeded far enough for the average level of prices to be determined by market demand. Under the legislative and regulatory regime in force from 1st January 1985, price-controlled gas will slowly work its way out of the supply system. New production from fields in these categories would be unlikely to replace existing old gas because of the economic constraints outlined above. The likely share of such gas in total interstate pipeline supplies in the early 1990s could be around 20% if this legislative and regulatory environment were to continue. Market pressures would tend, under the economic and energy price scenarios described in this study, to keep the average price of gas above the controlled old gas prices. Figure V-6c shows the implied relative prices and volumes of old and new gas under these conditions for any given average price of gas.

Complete deregulation of old gas prices would lead to these prices being bid up to the market wellhead price. Rolled-in pricing for other, higher cost gas would no longer be feasible — with the probable effect that less of these higher cost supplies would be forthcoming in the short run. Figure V-6d illustrates the effect on relative volumes of old and new gas. In this case there would, of course, no longer be any distinction between the two in terms of average prices. The substitution of more low-cost gas for high-cost gas could take place under deregulation, with the higher-cost gas reserved for later use as market pressures require. In the case shown in Figure V-6c by contrast, the old gas would remain forever unavailable as supply.

(iii) Expected and Possible Future Supply

Depending on the direction of legislative or regulatory change and on demand, supply from conventional gas resources in the lower 48 could rise from the 1984 level of 454 bcm (16 TCF) to 530 bcm (18.7 TCF), or could decline further to around 400 bcm (14 TCF). In the event of low production of old gas and continuance of rolled-in pricing, higher volumes of gas from tight sand formations could be expected, possibly up to 28 bcm (1 TCF), almost twice the 1984 volume. With complete price deregulation in place by 1990, the volume of tight sands gas economically recoverable could fall on the other hand to around 8 bcm (0.3 TCF). After 1990, conventional lower 48 gas supply, including all offshore production, is likely to be in long run decline. Table V-7 summarises the supply outlook for the United States.

The volume of Canadian gas exports will have an important impact on United States conventional and unconventional gas supply in the medium term. Canadian gas is also likely to be an important long-term

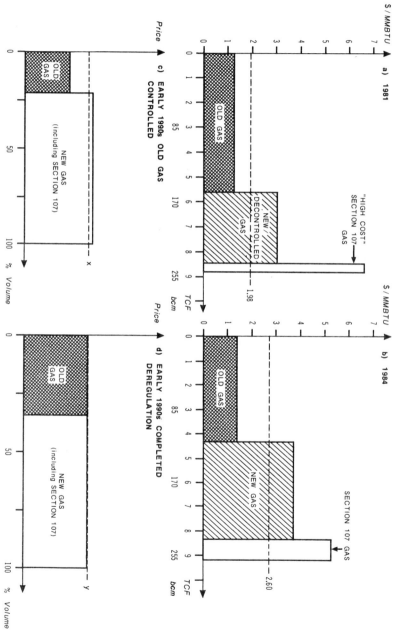

Figure V-6

Illustrative Effect of Complete Deregulation on U.S. Gas Supply and Prices to Major Pipelines in Interstate Market

Source: IEA Secretariat.

supply source for key United States markets in California, the midwest, and (assuming the approval of at least one of the current active proposals for major new export/import pipelines) to the Northeast — directly or by sales displacement from Gulf Coast supplies to the Midwest. The conventional gas producing areas of Canada are the Western Canada Sedimentary Basin, located in the provinces of Alberta, British Columbia and Saskatchewan, and parts of Eastern Canada. Western Canada produces more than 99% of the current total production of 73 bcm. These conventional producing regions are expected to be able to supply the increasing level of Canadian demand until well beyond the year 2000 given a reasonable range of reserve additions. By 2010,

Table V-7
Natural Gas Supplies for the United States
(billion cubic metres)

	1973	1985[1]	1990	2000	2010
Lower 48 States					
Conventional	590.6	459.8	400-530	340-450	230-340
Tight Sands	-	18.0	8- 28	10- 50	20- 75
Alaska	-	-	-	0- 20	20- 40
Sub-Total	590.6	477.8	408-558	350-520	270-455
Contracted imports					
Canada[2]	27.2	25.5	13	-	-
Mexico[3]	negligible	-	-	-	-
LNG	negligible	0.7	-	-	-
Sub-Total	27.2	26.2	13	-	-
Possible renewals and extension					
Canada			17- 47	20- 45	10- 40
Mexico			0- 10	5- 20	5- 40
LNG			0- 1	6- 10	6- 35
Sub-Total			17- 58	31- 75	21-115
Total	617.8	504.0	438-629	381-595	291-570
Total (TCF)	21.8	17.8	15.5-22.2	13.4-21.0	10.3-20.1
Of Which:					
United States lower 48 States					
Conventional (TCF)	20.9	16.2	14.1-18.7	12.0-15.9	8.1-12.0

1. Provisional and partly estimated.
2. Volumes with National Energy Board licence.
3. An operating contract to import gas from Mexico is currently suspended by mutual agreement of sellers and buyers, but could be reactivated at any time.

incremental supplies from low permeability ("tight sands") reservoirs or more remote natural gas deposits (in the McKenzie Delta, Arctic Islands or offshore regions) may be required. Sufficient economically recoverable gas is available for the Canadian Government to have authorised large-scale exports of gas to the United States and Japan. Current authorisations permit exports to the late 1990s and in some cases to the year 2000.

In the medium term, some increase in Canadian export sales to lost traditional markets can be expected, since the adoption by the Canadian authorities of the policy that exports will be permitted as long as the price of exported gas is not lower than the price paid by Canadian consumers in areas or zones adjacent to the exporting region. Through to 1990, Canadian exports could be in the range of 20 to 50 bcm, depending on locally competitive conditions and on competition from alternative United States sources. At the top end of the range, Canadian supplies in the medium term could depress conventional lower 48 United States production below its sustainable capacity, prolonging the so-called gas "bubble" through to 1990.

A high rate of production of Canadian reserves in the medium term will tend, however, to have one of two effects. If prices in final markets remain relatively low, reserves replacement ratios can be expected to be low as well, and longer term exports could be limited by concern about the possibility of supplying the Canadian domestic market with gas. If, on the other hand, price incentives were to lead more Canadian resources to be converted into active reserves, then the same price signals would tend to make imports to the United States from further afield relatively more attractive, and to stimulate frontier technology for the development of more gas from tight sands and less conventional sources in the United States itself.

There are widely divergent views about the long-term geological prospects for both lower 48 and Canadian conventional gas production. In general, many major gas producers have a pessimistic outlook (which would tend to the bottom end of the ranges shown in Table V-8). If this outlook is confirmed, it could lead to a rise in gas prices substantial enough to promote non-conventional resources, large-scale imports, or the development of a delivery system for Alaskan gas to the lower 48. The alternative would be appreciably lower demand and an increased need for oil imports. Certainly by the year 2000, one or more of these options is likely to be in force. Table V-7 identifies possible ranges for volumes of gas from these sources.

Potential sources of imports, in addition to Canada, are Mexico and existing or future possible LNG suppliers. Mexico delivered about 3 bcm (0.1 TCF) of gas by pipeline to the United States for three years 1980 through to 1982. The volume of deliveries declined steadily through 1983 and 1984 as more competitive domestic United States supplies became available. Deliveries of Mexican gas to the United States were suspended indefinitely late in 1984. No resumption is anticipated in the short term, but the resource base of Mexican gas appears adequate to support exports to the United States market at some future date without impeding supplies to Mexico's own domestic market. Proven reserves are estimated at about 2 200 bcm (77 TCF), with additional resources of over 5 000 bcm (175 TCF). However, the possibility of exports may be constrained by lack of capital to construct a large delivery system for Mexican gas to the United States border.

There are four LNG receiving terminals in the United States, in Massachusetts, Maryland, Georgia and Louisiana, although only one has been operating in recent years and maintenance work would be required to enable the others to operate. Spare liquefaction capacity and ships are available for delivery of LNG from Algeria, and imports of up to 6 bcm per annum could be delivered from this source without construction of new facilities either in Algeria or the United States. Imports will depend on commercial terms and on approval by the Energy Regulatory Administration. Future LNG sources for the United States could include Western Hemisphere sources, Norway and Indonesia. If a project were to be developed in Nigeria to supply European markets, incremental liquefaction capacity could potentially also be constructed to meet emerging United States needs around the turn of the century or thereafter.

The price and demand paths used in this study imply that by 2000, under a fully market-price regime, Mexican and LNG imports are again likely to be supplying the United States. It is probable that such imports would be required on a larger scale than in the late 1970s, although in the initial stages imports from these sources could balance small disequilibria in the supply and demand of United States and Canadian gas. Four or five world scale LNG export projects would be required to provide the supply of LNG shown at the top end of the range for the year 2010.

The prospects for Alaskan gas are less certain in a market where prices could respond quickly to gas-on-gas competition, because large volumes would have to be asorbed quickly and securely by the market to justify the very high capital cost of a pipeline transmission system to the lower

48 states. Reserves of about 1 000 bcm (35 TCF) are regarded as proven by the Alaskan Oil and Gas Conservation Commission, with an additional 3 900 bcm (137 TCF) believed to exist in the onshore and offshore areas of the state. Over 85% of reserves are located on the North Slope, with no means of transporting the gas to market. The pipeline project to deliver the gas to the lower 48 states, the Alaskan Natural Gas Transportation System, received certain important regulatory approvals, but financial support has not been forthcoming because of market conditions. It is likely that energy prices in general and gas prices in particular would have to increase substantially before companies are willing to take the risk exposure which construction of ANGTS would involve. The project is unlikely to be revived until the 1990s. Alaskan, especially North Slope, gas may nevertheless be an important new source of supply in the next century.

Figure V-7 compares the possible supply range with the demand range for the United States given in Appendix II, Table 2. A wide range of

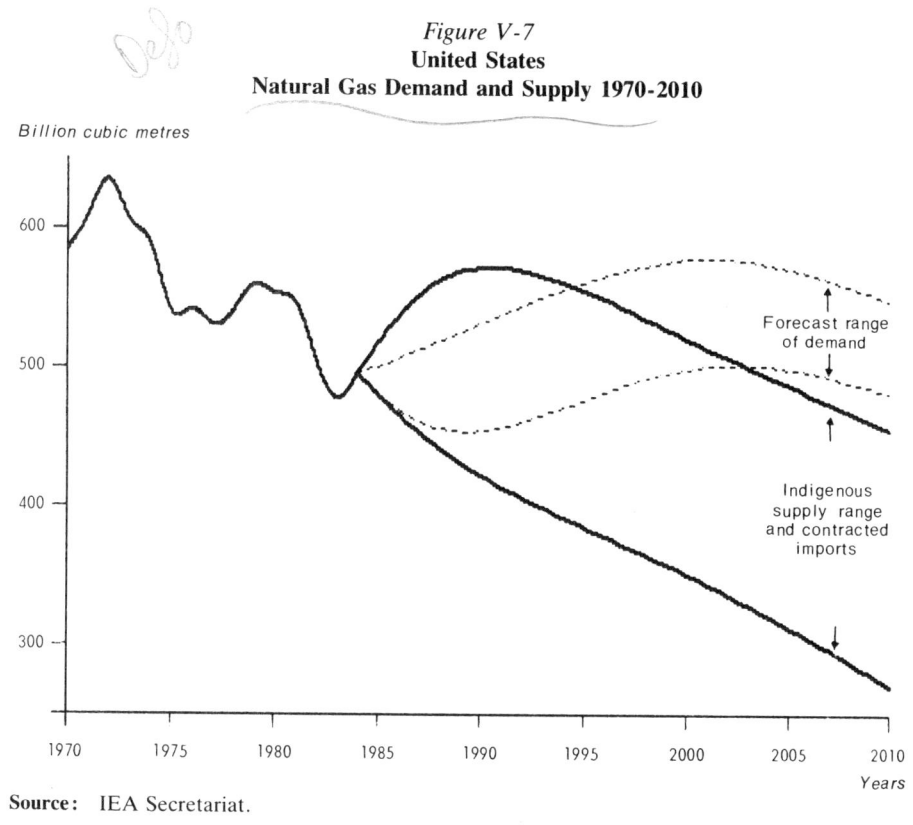

Figure V-7
United States
Natural Gas Demand and Supply 1970-2010

Source: IEA Secretariat.

speculative, non-contracted supplies are included already before 1990. This contrasts with the profile of gas supply and demand for OECD Europe (Figure V-4), where most of the supply is firmly contracted until the mid-1990s. As a result of declining conventional lower 48 reserves and of the changes in market organisation to a much more short-term oriented supply-demand response, the transmission companies in fact cover relatively little forward demand. Figure V-8 shows the expected delivery profile of 38 major interstate pipelines, as reported to the Federal Energy Regulatory Commission for the 20 years from 1984 through to 2003, based on contracted supplies. For comparison, Figure V-8 also shows the same companies' expected delivery profile out of contracted supplies as reported in 1964 for the 20 years to end 1983. The sharply declining curve based on 1984 expectations contrasts strongly with the more stable expectations of 1964, and also with slower rate of decline of firmly committed contractual cover in Europe. This gives a measure of the increase in recent years of the exposure of the United States gas industry to market dynamics and their related risks.

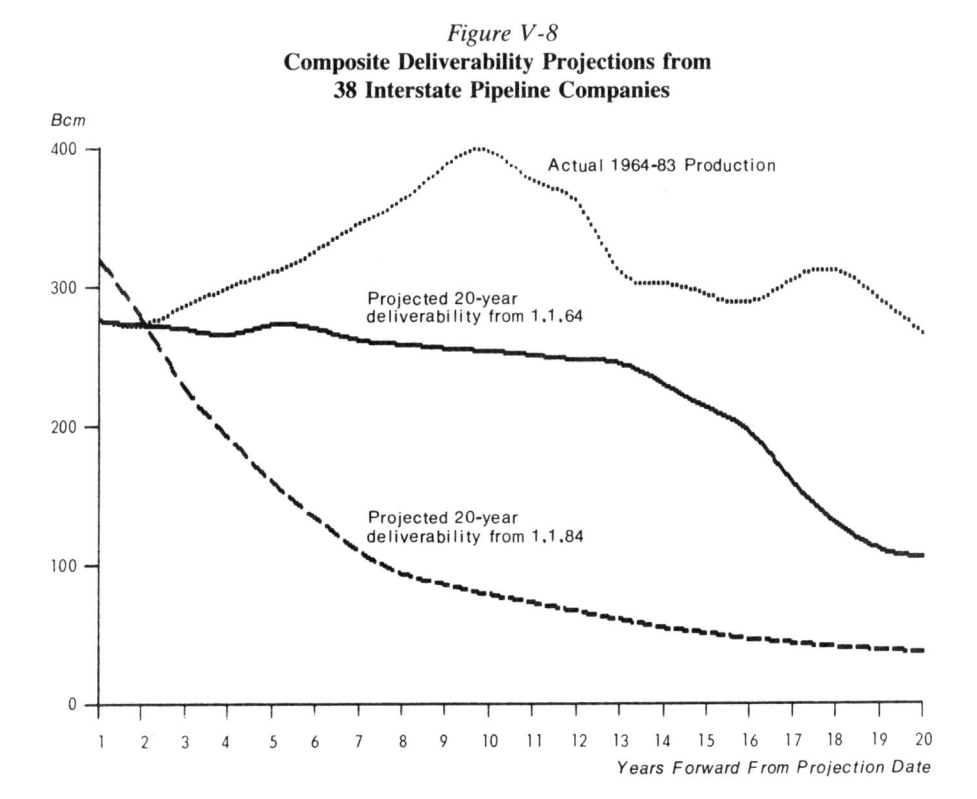

Figure V-8
**Composite Deliverability Projections from
38 Interstate Pipeline Companies**

Source: Based on figures from the from the United States' Energy Information Administration.

E. Supply Options for Japan

Japan's small quantities of domestic natural gas are supplemented by imports of LNG largely from developing countries outside the OECD. There are three major suppliers (Brunei, Indonesia from two separate production facilities, and Malaysia) and two minor suppliers (United States, from Alaska, and the United Arab Emirates, from Abu Dhabi). Each of these contracts will probably run to the year 2000 and beyond; where reserves and facilities for liquefaction and transportation of gas are in place, extension of some of these contracts is possible. Some supplies over and above average contracted volumes of gas have been made available to Japanese buyers as needs arise, and it is likely that this will continue to be a feature of the supply picture in the future. New supplies will be available by 1990 from Australia.

Table V-8 summarises the existing contracts and those in an advanced stage of preparation. The forecasts show average annual contract

Table V-8
Natural Gas Supply to Japan: Contract Volumes 1983-2010
(Million Metric Tons of LNG)

	1983	1984	1990	2000	2010
Contracted					
Alaska	1.0	1.0	-	-	-
Brunei	5.2	5.2	5.2	-	-
Abu Dhabi	1.8	2.1	2.1	-	-
Indonesia	10.6	13.5	14.7	14.0	-
Malaysia	1.8	3.4	6.0	6.0	-
Australia	-	-	2.7	5.9	-
Sub-Total	20.4	25.3	30.7	25.9	-
Possible Extensions					
Alaska			1.0	1.0	1.0
Brunei				5.2	5.2
Abu Dhabi				2.1	2.1
Indonesia				-	14.0
Malaysia				-	6.0
Australia				-	5.9
Sub-Total			1.0	8.3	34.2
Total			31.7	34.2	34.2

Source: IEA Secretariat.

volumes; some flexibility is available to meet actual needs on a year-by-year basis. Other LNG projects are proposed by various interests in Qatar, Thailand, the Soviet Union (the Sakhalin Project) and the United States (Alaskan gas from the Cook Inlet and the North Slope) as well as in another region of Australia (the Bonaparte Gulf). It is, however, highly probable that average annual contract volumes under existing contracts and domestic production would be sufficient to cover the demand forecast in 1990 (see Table V-9), and could be sufficient through to the year 2000 if renewals or extensions are arranged equivalent in volume to the existing Brunei and Abu Dhabi contracts (due to expire in 1992 and 1997, respectively).

Table V-9
Natural Gas Supply/Demand Balance 1983-2010
Japan
(billion cubic metres)

	1983	1984	1990	2000	2010
Demand	30.2	38.6	45.0-53.4	51.5-66.0	56.0-72.0
Supply from:					
Domestic Production	2.2	3.0	4.0- 4.5	4.7- 6.3	3.1- 4.8
Required LNG Imports (bcm)	28.0	35.6	41.0-48.9	46.8-59.7	52.9-67.2
Required LNG Imports[1] (million tons of LNG)	20.4	25.9	29.8-35.6	34.0-43.4	38.5-48.9
Contracted LNG Imports (million tons of LNG)	20.4	26.6[2]	30.7	25.9	-
Surplus/(Deficit)	-	0.7	0.9-(4.9)	(8.1)-(17.5)	(38.5)-(48.9)
Possible extensions to LNG Contracts and new projects[2]			1.0	8.3	34.2
Balance to meet demand			0-3.9	0-9.2	4.3-14.7

1. Assuming 1.375 bcm equals 1 million metric tons of LNG.
2. Actual volumes imported in 1984 slightly exceeded contracted volumes (which are shown in Table V-8).

Source: IEA Secretariat.

VI. THE ENVIRONMENTAL EFFECTS OF NATURAL GAS PRODUCTION, TRANSPORT AND USE

A. Exploration, Production and Transport of Gas

The environmental impacts of gas exploration and production of gas are very similar to the impacts from oil exploration and production. The impacts include discharges of drilling mud and of oil infested water. These are in particular environmental hazards connected with offshore drilling. The risk of pollution from leakages is, however, most serious for oil production and transport. Leakages from gas production and transport pose less of an environmental problem but are of course a hazard because of the risk of fire or explosion. There is in general a higher risk of blow-outs from gas wells than from oil wells with a resulting higher risk of fatal accidents in gas exploration and production than in oil exploration and production.

Natural gas is found as sour gas in many reservoirs, that is, it contains sulphur components like hydrogen sulphide (H_2S). Before being piped into the transmission system the gas is cleaned of sulphur components. The existing technologies allow a very high degree of cleaning with a removal of up to 99% of the pollutants. This high degree of cleaning is made necessary by the acidity of the sulphur components, which would otherwise pose serious corrosion problems in the pipeline system. The choice of technology for removing sulphur components depend on the composition of the gas as well as on the characteristics of the field. The removed H_2S is a toxic gas which is not tolerated as an emission. In some instances H_2S may be converted to sulphur dioxide (SO_2) and emitted to the atmosphere. It is, however, in many cases not allowed and the H_2S

removed must be converted to elemental sulphur or some other form to prevent its release to the atmosphere. The sulphur recovery efficiency of the most common utilised technology has continuously improved.

Pipeline transport of gas has no significant environmental impacts. Most pipelines are buried and the impact on vegetation of a pipe-laying operation will in many circumstances be minimal a few years after the pipes are laid. The major exception is pipelines crossing Arctic regions where burying is excluded and the construction creates a corridor of destroyed vegetation, which is difficult to restore. Transmission lines are carefully tested after laying for possible leaks through fractures or welding and are when in use normally inspected and tested regularly. The risks of leaks are mainly confined to the medium and low pressure distribution grid, which is more exposed to damage by roadworks or other construction activity or by settling of the ground as a result of, for example, heavy road traffic. Furthermore, part of the distribution grid is relatively old. The old pipes are more fragile than modern pipes, which very often are made of polyethylene and which are more flexible and have fewer joints than steel pipes or the cast iron pipes used earlier in the town gas distribution systems. Methods for internal inspections of pipes have been developed together with safe methods to repair leaks by sealing which in many circumstances is less costly than replacing the existing pipes.

Gas may also be transported in liquefied form, as LNG, which is transported and stored under atmospheric pressure at very low temperatures. Safety requirements are such that spillages or accidents have a low probability, but the impact on the nearby environment of an accident resulting in an explosion could be serious. This requires precautions when choosing the siting of LNG facilities. Special measures for navigational safety are equally necessary in order to minimise the risk of collision of LNG carriers with other ships. LNG storage tanks — both on carriers and on land — are usually double walled, although some older single-walled tanks are still in service in some parts of the world. The inner wall requires a material with good cryogenic properties, i.e. the ability to withstand stress at low temperatures. For this purpose most inner walls are built of special steel, of concrete, or of aluminium. The outer wall is normally a carbon steel shell, but shells of concrete have also been constructed in order to increase the resistance of the tanks against mechanical shocks, for example provoked by an air crash. Best industry practice also currently includes earth-construction surrounds to the concrete shells in order to resist direct impact mechanical shocks.

B. End Use of Natural Gas

Safety requirements for new appliances are high and special safety features are incorporated in the equipment, but the risk of malfunctioning increases with the age of the equipment. The risk of malfunctioning can be minimised by maintenance and control of the equipment.

Table VI-1
Emission Factors for Industrial Boilers before Control Equipment
(kilograms of emissions per toe of fuel use)

	Coal (1% S 10% ash)	Fuel Oil (1% S)	Natural Gas
Particulates	100.0	1.8	0.1-0.3
Sulphur Oxides	29.2	20.0	0
Carbon Monoxide	1.5	0.7	0.3
Hydrocarbons	1.5	0.1	0
Nitrogen Oxide	11.5	8.2	2.3-4.3

Source: United States Environmental Protection Agency, AP-42 "Compilation of Air Pollutant Emission Factors".

Table VI-2
Emission Factors for Oil and Gas Use in the Residential Sector
(kilograms of emissions per toe of fuel use)

	Gas Oil	Natural Gas
Particulates	0.4	0.1-0.3
Sulphur Oxides	5.9	0
Carbon Monoxide	0.7	0.15
Hydrocarbons	0.1	0.1
Nitrogen Oxides	2.6	1.6

Source: United States Environmental Protection Agency, AP-42 "Compilation of Air Pollutant Emission Factors".

The cleaning of the gas where necessary at the production phase implies that the product received by the customer can be burnt in a manner which has only minimal impacts on the atmosphere and considerably less than the burning of other fuels normally will have. This is illustrated in Tables VI-1 and VI-2, where air emissions from the burning of alternative fossil fuels in industry and households are compared. The most important emissions from gas combustion are nitrogen oxides, which are formed in the combustion process by the oxidisation of the nitrogen present in air, but these can be kept at a relatively low level by appropriate burning controls and techniques. In the combustion of oil and coal also the nitrogen contained in the fuels contribute to the emissions of nitrogen oxides. Nitrogen oxides formation from the nitrogen content of the air is a function of the temperature in the combustion chamber and the rate of cooling of the combustion products. Emission levels vary considerably with the type and size of unit and are also a function of the load on the boiler. Several operating modifications can be employed as staged or two-stage combustion which contribute to lower nitrogen oxides emissions. Other possibilities include low excess air firing and flue gas recirculation. Low excess air firing can reduce nitrogen oxides formation by 10 to 30% without producing unacceptable levels of unburned combustibles like carbon monoxide, hydrocarbons and smoke. For large utility boilers flue gas cleaning by a denitrification process is feasible. Research and development is continuing in burner designs having the prospects of allowing further reductions of nitrogen oxides emissions. In most cases such new boiler designs also involves a higher efficiency. In the particular case of using natural gas in compressed form as a motor vehicle fuel (CNG), nitrogen oxide emissions tend to be high, and this tends to act as an environmental constraint on the acceptability of this technology for using gas.

C. Conclusion

Gas is delivered to the customer as a clean fuel. The use of natural gas has, compared to other fossil fuels, environmental advantages and substantial improvements in local air quality have occurred in urban areas where gas has substituted for other fuels. The major air-pollutant from gas burning is nitrogen oxides but emissions of nitrogen oxides are generally much lower from gas burning than from burning of oil or coal and can be reduced more readily.

In the production phase, gas well drilling and operation has environmental impacts very much like that of oil except for the effects of leakages, which can be more serious in the case of oil compared to gas. Gas pipelines are usually buried and have practically no environmental impacts apart from problems of noise from compressor stations and from pressure reduction stations. These impacts are minimal compared with the effects of surface transportation of oil and coal products.

VII. STORAGE AND CONTINGENCY PLANNING

A. Seasonality and Security

Stocks of gas have a dual role in ensuring that supply matches demand. First, transmission and distribution companies must hold gas in stock at the beginning of the season of peak demand to balance the rate of deliverability from suppliers with the rate of demand from consumers. In performing this load balancing function, storage of gas is one element among many others — seasonally interruptible supplies and flexible offtake from suppliers are the main alternatives. The amount of gas storage in place should be determined in a market economy by the relative economics of constructing and operating storage against the cost of these other means of balancing load such as interruptible customers in industry or electricity and flexibility in supply. Secondly, storage of gas can act as an emergency reserve in the event of shortages in supply, for whatever reason they may occur. This may be a particularly important function of storage for countries and companies which depend heavily on imports of gas. Interruptible sales contracts, the possibility of switching supplies between companies, regions and countries and a diversity of suppliers also enhances security of supply.

Demand for gas in peak seasons corresponds closely to variations in temperature, especially to temperature extremes. The capacity of storage systems, as well as of transmission networks, is designed by gas utilities to cover a defined probability of cold weather. Typical cover among European utilities aims to guarantee supply (at a national level) of a volume of gas adequate to meet the requirements of a severe winter likely to occur one year in 40 or 50. In the case of some countries which rely heavily on imports, this criterion may be reinforced by an assumption that in that same winter, the modulation of supplies could be inverted (to give a summer peak in supply).

There will always be a trade off between the amount of storage needed for load-balancing and seasonal peak cover and the amount available as a contingency reserve. In the extreme case of a very severe winter and inverse supply modulation, existing capacity could be expected to be almost fully utilised in most West European countries. In a "normal" winter, for continental European importers, a rough estimate would be that 50 to 60% of available working gas in storage could be expected to be used for load balancing needs, with the balance available as a "pure" contingency reserve. The existence of extra physical supply capacity at the producing end, especially in the Netherlands but also in domestic production in Germany and Italy must also be taken into account as a contingent source of supply, although there are not always formal arrangements in place for this to be made available over a sustained period.

Table VII-1 shows existing gas storage capacity at 1985 in IEA importing countries and expected capacity in 1990 based on work in progress. Volumes shown are working gas volumes, i.e. excluding the gas cushion which is required in some storage facilities to maintain pressure. Four principal types of storage have been included in the table — aquifers, depleted gas fields, salt caverns and LNG tanks (both receiving terminals and peak-shaving facilities). The main use of salt cavern storage in some countries is for gas quality control. The main use of LNG storage in most countries (Japan and Spain are the exceptions) is for meeting peak demands — LNG typically is expensive per unit of storage capacity provided, but the cost per unit of output capacity is low. All forms of storage are not necessarily equal in economic or technical terms for different load-balancing and security functions. Table VII-2 illustrates the relative importance of cavern and LNG storage in the main European importing countries.

It is clear from Table VII-1 that the majority of IEA gas storage is in the United States. Against annual consumption needs of around 500 bcm, the United States has storage capacity of 234 bcm. In Europe, against consumption of just over 200 bcm, there is storage capacity of 25 bcm, rising to about 33 bcm by 1990. In Japan, consumption of 30 bcm (likely to increase to about 50 bcm by 1990) is covered by storage of under 5 bcm. There are important institutional, geographical and market reasons for these differences.

Table VII-1
Gas Storage Facilities in Selected IEA Countries

Country	Working Gas Capacity (bcm)		Maximum Daily Offtake (Million cubic metres per day)	
	1985	1990	1985	1990
Austria	2.03	2.20	14.1	19.2
Belgium	0.48	0.63	18.8	40.8
Denmark	0.04	0.23	4.8	7.2
Germany	3.67	6.99	102.5	118.6[1]
Italy	9.00	11.03	135.0	191.0
Spain	0.14	0.19	11.2	15.4
Sweden[2]	-	-	-	-
Switzerland[2]	-	-	-	-
United Kingdom	2.74	2.88	75.0[3]	75.0[3]
IEA Europe	18.1	24.15	361.4	467.2
Japan	4.1	5.8	97.6	138.0
United States	234.0	n.a.	2 040.0	n.a.
IEA TOTAL	256.2	n.a.	2 499.0	n.a.

1. Partly estimated.
2. Exploration work proceeding for suitable storage reservoirs.
3. Expandable to 115.0 Million cubic metres per day in emergency.

Table VII-2
Storage Types in Selected IEA Importing Countries

Country	Storage Capacity			
	As % of Total Working Gas		As % of Maximum Daily Offtake	
	Aquifers and Depleted Gas Fields	Caverns	LNG	LNG
Austria	100	0	0	0
Belgium	87.5[1]	0	12.5	4.5
Germany	74	25	1	3 (est.)
Italy	99	0	1	4
Spain	0	0	100	100
United Kingdom	82	3	15	53

1. Including disused coal mines.

In Japan since natural gas is used primarily in power generation, for baseload moving to middle load supply, the seasonal swing in consumption is very moderate. Indeed there is a slight summer peak in this sector of consumption (because of electricity demand from air conditioning) which offsets the small winter peak in the space heating markets. Security needs are covered by the ability of LNG-fired power stations to switch to fuel oil or even to mix fuel oil in the burner racks in the event of LNG supply difficulties.

The difference in European and United States needs for storage can be seen by comparing Figures VII-1 and VII-2. These figures superimpose, for a typical heating season, the seasonality of supply and consumption in the United States (Figure VII-1) and Western Europe (Figure VII-2). In the United States, the ratio of production in any given month to average production through the year is fairly stable around a value of 1.0, with only a slight mid-winter peak in January. The profile of consumption, as expected, is highly variable and the difference between the two is met essentially by drawing on storage in the heating season and filling during the summer months. In Europe, by contrast, both production and consumption are highly variable according to season. Less storage capacity is therefore required to meet seasonal demand.

The explanation for the vary large differences in the way in which supply and demand loads are matched in the United States and Europe is partly geographical. In the United States, the major markets of the Midwest and northeastern states and of California are far removed from the major southwest central producing states. In Continental Europe, the large markets of the Ruhr region, the Low Countries and northern France are relatively closely situated to the Netherlands gas province on which the rapid growth of the industry has been built.

There is also an important historical institutional explanation. In the United States, transmission companies' rates have been for many decades regulated by state and federal authorities with rates normally determined to give a fixed percentage rate of return on assets. One of the determining factors in setting rates is the valuation of company assets allowable in the "rate base". Under such a system, transmission companies could increase the absolute value of their earnings by demonstrating the need to make investments which would increase their underlying asset valuation. Storage facilities have normally been allowable in the rate base. The incentive to invest in storage facilities has therefore been strong. At the same time, during the period of universal price controls on the wellhead price of natural gas destined for interstate sales, producers were reluctant to invest in field capacity which might lie idle for part of the year. The existing configuration of storage and swing production capacity were thus distorted by institutional features and does not necessarily bear any relation to the costs of balancing load by these two means.

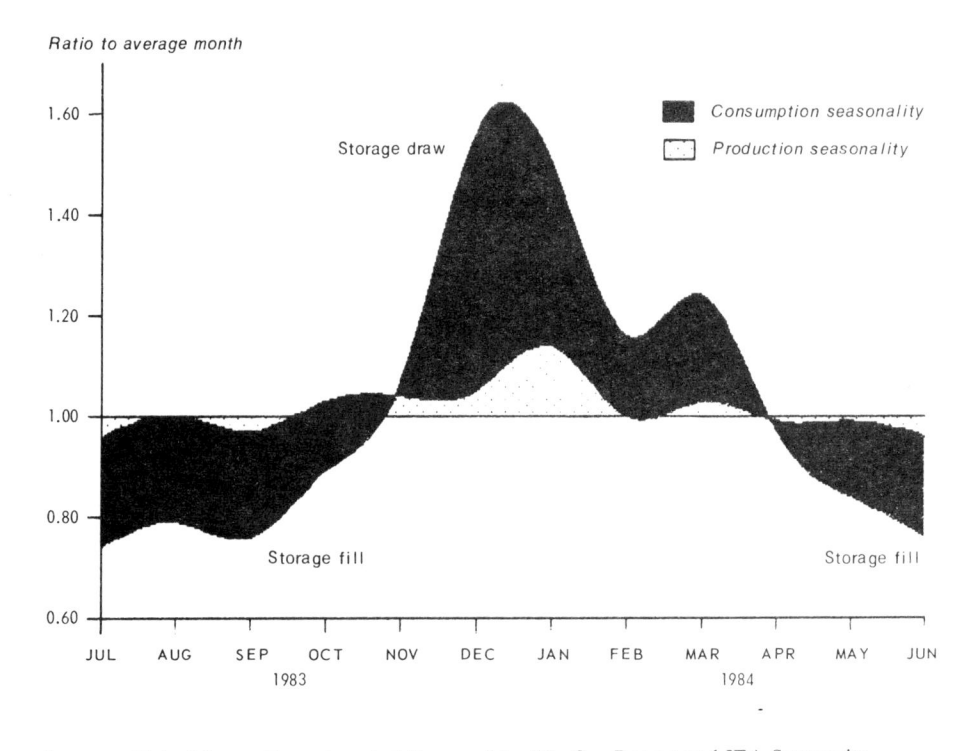

Figure VII-1
United States Gas Seasonality
Heating Season 1983-84

Source: United States Department of Energy, Monthly Gas Report and IEA Secretariat.

In Europe no such institutional distortions of the market existed. Indeed, the degree of integration, overlap and communication between companies at the production and transmission stages of gas supply may have tended to lead to a concern for optimisation of the costs of the overall system. Storage can in principle be provided more cheaply from onshore

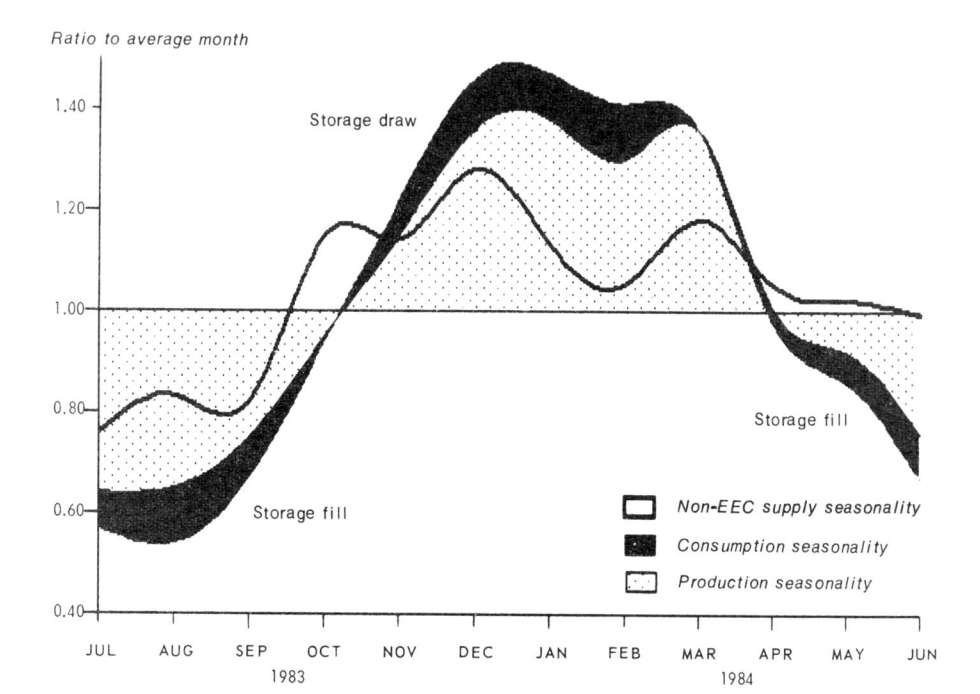

Figure VII-2
European Gas Seasonality
Heating Season 1983-84

Ratio to average month

Storage draw

Storage fill

Storage fill

□ *Non-EEC supply seasonality*

■ *Consumption seasonality*

▨ *Production seasonality*

JUL   AUG   SEP   OCT   NOV   DEC   JAN   FEB   MAR   APR   MAY   JUN
1983   1984

Source: IEA Secretariat.

production facilities if the reservoir characteristics are suitable than from purpose-built aquifer, depleted field or cavern storages — notably because there is no requirement to purchase and inject cushion gas (which is then lost) simply to maintain pressure. Where offshore fields or difficult reservoirs are concerned then the relative economics may be more marginal. The unique role of the Netherlands Groningen field, and the United Kingdom North Sea Southern Basin fields in providing high load factors of supply has meant that relatively little market pressure to build extensive storage cover has been required. However, for countries and regions more remote from such substantial swing suppliers (such as Austria, Italy and parts of France), storage has played and continues to play a more important role in covering seasonal demand.

The United States gas market has contracted substantially in volume terms since the peak year of 1973 and is unlikely, according to the forecasts, to recover to exceed that level. For the future, storage capacity should therefore remain more than adequate.

As for Western Europe, the most interesting future prospects probably concern this question of load balancing capacity. On the one hand, the growth of residential sector and other space heating demand relative to bulk uses is likely to lead to increased seasonality of demand. Storage expansion schemes are already in progress in most European countries in anticipation of this worsening demand load factor (Table VII-1) in the medium term. On the other hand, significant long-term changes in supply load factors could result from declining production in the large, flexible Southern Basin fields in the United Kingdom North Sea and from the expected fall off in pressure in the Groningen field in the Netherlands. The rate at which the Groningen pressure-drop occurs is likely to become sharpest in the first decade of the next century.

B. Conclusions

In order to maintain the same degree of security of supply, given the projected trends in supply and demand load factors, increased attention will have to be paid to the question of ensuring that seasonal supplies and demands are well matched. This may require careful attention to market trends as well as ensuring that price developments do not unduly damage the prospects for gas' competitiveness in industrial baseload markets especially in the interruptible section of these markets where dual-firing capabilities and alternative stocks held on a commercial basis enable customers themselves to make a contribution to the overall security of supply as well as to load-balancing. There is likely to be a requirement either for considerable seasonal flexibility in the deliverability profile of the next major tranche of gas in Western Europe or for further expansion of storage systems.

If adequate flexibility were not forthcoming in future tranches of gas supply, other arrangements to ensure security would need to be enhanced. Improved regional co-operation and the physical flexibility to move gas from different sources around the European transmission grid would be appropriate. There have been no major changes along these lines in Europe in recent years, although in northern Italy the expansion of grid capacity linking key storage fields will enable most Italian markets to be served effectively by a variety of suppliers. However, the

Dutch commitment to extend existing export contracts beyond their original expiry dates in the 1990s has increased the proportion of gas supply available on a relatively flexible basis. Accordingly, better supply security is now assured in Europe for a given level of storage capacity in future years than appeared to be the case a few years ago.

Technical Assumptions on Energy Prices and Economic Growth Rates

The forecasting range is based on two sets of assumptions relating crude oil prices and economic growth. The assumed paths for oil prices and economic growth are shown in Appendix Table I-1. In both cases oil prices are assumed to decline in real terms between 1985 and 1990, then to rise on average for the rest of the century and beyond; in the high oil price case they are assumed to increase more rapidly after 1990. For the OECD as a whole, economic growth rates are assumed to average 3.5% per annum from 1983 to 2010 (oil prices $20.50 in 1990 and $30.00 in 2000) and 2.5% per annum (oil price $22.00 in 1990 and $45.00 in 2000; all prices in constant 1984 dollars) in the high and low growth cases, respectively. Exchange rates for the United States dollar against other currencies are for the analysis assumed to stabilize at around their 1983 levels, i.e. below the levels reached in 1984 and early 1985. Neither outlook has considered significant changes to energy policies.

Substitution between fuels takes place over time in response to changing fuel price relationships in each sector of consumption. The relative fuel price assumptions in this study are set out below:

- Oil product prices maintain their end-1985 relationship with crude oil prices.

- Average coal prices remain constant in real terms through 1995, then increase at an average of 2% per year. Domestic coal

Appendix I Table 1

Crude Oil Price and Economic Growth Scenarios

High Growth/Low Oil Prices
Constant 1984 $/bbl 34° Crude
FOB Middle East

		Annual Average Per Cent Change in GDP			
			Europe	N. America	Pacific	TOTAL OECD
1983	29.00				
1984	28.50				
1985	26.90	1983-1990:	2.9	3.5	5.0	} 3.5
1990	20.50				
1995	24.80	1990-2010:	3.3	3.3	4.7	} 3.5
2000	30.00				
2010	48.00				

Low Growth/High Oil prices
Constant 1984 $/bbl 34° Crude
FOB Middle East

		Annual Average Per Cent Change in GDP			
			Europe	N. America	Pacific	TOTAL OECD
1983	29.00				
1984	28.50				
1985	26.90	1983-1990:	2.0	2.1	4.1	} 2.5
1990	22.00				
1995	31.35	1990-2010:	2.3	2.3	4.0	} 2.5
2000	45.00				
2010	55.00				

production is expected to remain in use, in some OECD countries, and the price of imported coal is expected to reflect slightly increasing production costs.

- Electricity prices are assumed to remain constant in real terms to 1995, and then to increase at an average of 3% per year as input fuel costs increase. Gas prices to final consumers are assumed to remain in the same relationship to oil prices which they currently have in each of the main consuming regions of the OECD. Note that the price relationships between gas and oil remain different throughout the forecasting period for each of the main OECD consuming areas. This reflects different institutional arrangements in Europe, Japan and the United States and the assumption of no possibility for arbitrage between markets.

It should be stressed that these forecast relationships are working assumptions. It is also possible, for example, that:

- There may be some change over the long-term in the current relationship between oil and gas prices to the final consumer.

- There may be inter-connections between different regional markets if, in the long-term, common suppliers are involved with more than one region.

- Structural changes in relative oil product prices as a result of changes in oil markets, the refining industry, and oil product trade patterns, could invalidate the assumed constancy of the relationship betwen gas prices and certain oil product prices.

- Coal prices could be subject directionally to pressure from oil and gas prices; and

- Electricity prices may be vulnerable to interest rate changes and their impact on the servicing of capital costs.

APPENDIX II

Natural Gas Demand Outlook by Sector of Consumption

The tables in this Appendix show the outlook for natural gas demand by main sector of consumption as calculated by the IEA Secretariat using the assumptions outlined in Appendix I. A descriptive analysis of demand prospects for each country is given in Chapter IV, section B.

Appendix II Table 1
Natural Gas Demand by Sector: 1973-2010 OECD Europe
(billion cubic metres)

	1973	1984	1990	2000	2010
Low Demand/High Oil Price					
Industry	57.8	77.0	78	84	91
Residential/Commercial	49.3	101.0	115	130	139
Power Generation	26.7	25.4	22	23	17
Transport	0.0	0.4	0.5	0.5	0.5
Own Use and Losses	6.5	8.2	9.5	10.5	10.5
TOTAL (bcm)	140.3	212.0	224	248	258
TOTAL (Mtoe)	122.2	184.7	195	216	225
High Demand/Low Oil Price					
Industry	57.8	77.0	82	102	119
Residential/Commercial	49.3	101.0	120	133	148
Power Generation	26.7	25.4	32	32	25
Transport	0.0	0.4	0.5	1	1
Own Use and Losses	6.5	8.2	9.5	12	12
TOTAL (bcm)	140.3	212.0	244	280	305
TOTAL (Mtoe)	122.2	184.7	212	244	266

Natural Gas Demand by Sector: 1973-2010 United States
(billion cubic metres)

	1973	1984	1990	2000	2010
Low Demand/High Oil Price					
Industry	210.1	151.1	141	152	162
Residential/Commercial	204.0	193.6	210	228	245
Power Generation	98.2	85.0	43	57	20
Transport	-	-	1	3	3
Energy Sector[1] and					
Losses	92.4	59.2	59	59	57
TOTAL (bcm)	604.7	488.9	454	499	487
TOTAL (Mtoe)	526.9	425.9	395	434	424
TOTAL (Tcf)	21.4	17.3	16.0	17.6	17.2
High Demand/Low Oil Price					
Industry	210.1	151.1	157	187	185
Residential/Commercial	204.0	193.6	229	250	275
Power Generation	98.2	85.0	76	76	30
Transport	-	-	2	4	6
Energy Sector[1] and					
Losses	92.1	59.2	67	61	53
TOTAL (bcm)	604.7	488.9	531	578	549
TOTAL (Mtoe)	526.9	425.9	462	503	478
TOTAL (Tcf)	21.4	17.3	18.7	20.4	19.4

1. In the United States, significant quantities of natural gas are used in the oil extraction (enhanced oil recovery) and oil refining industries. These are included in the energy sector, not in industry, in line with historical practice in OECD statistics. The breakdown for 1983 was:

Oil and gas extraction	27.2 bcm
Oil refining	15.6 bcm
Pipeline fuel and loss	15.3 bcm
Total	58.1 bcm

Natural Gas Demand by Sector: 1973-2010 Canada
(billion cubic metres)

	1973	1984	1990	2000	2010
Low Demand/High Oil Price					
Industry	12.9	21.5	20	28	34
Residential/Commercial	14.0	22.4	23	30	32
Power Generation	4.9	1.7	1	2	2
Transport	-	-	-	-	-
Own Use and Losses	12.7	6.2	7	9	11
TOTAL (bcm)	44.5	51.8	51	69	79
TOTAL (Mtoe)	38.8	45.1	44	60	69
High Demand/Low Oil Price					
Industry	12.9	21.5	22	30	37
Residential/Commercial	14.0	22.4	26	36	39
Power Generation	4.9	1.7	2	2	2
Transport	-	-	-	1	2
Own Use and Losses	12.7	6.2	8	11	13
TOTAL (bcm)	44.5	51.8	58	80	93
TOTAL (Mtoe)	38.8	45.1	51	70	81

Natural Gas Demand by Sector: 1973-2010 Japan
(billion cubic metres)

	1973	1984	1990	2000	2010
Low Demand/High Oil Price					
Industry	1.9	3.3	4.0	6.0	8.0
Residential/Commercial	1.7	5.9	6.5	9.0	10.0
Power Generation	2.3	28.8	33.0	35.0	36.0
Transport	-	-	-	-	0.4
Own Use and Losses	0.2	0.6	1.5	1.5	1.6
TOTAL (bcm)	6.1	38.6	45.0	51.5	56.0
TOTAL (Mtoe)	5.3	33.6	39.2	44.9	48.8
High Demand/Low Oil Price					
Industry	1.9	3.3	4.5	7.0	10.0
Residential/Commercial	1.7	5.9	8.3	12.0	14.0
Power Generation	2.3	28.8	39.1	45.0	45.0
Transport	-	-	-	-	1.0
Own Use and Losses	0.2	0.6	1.5	2.0	2.0
TOTAL (bcm)	6.1	38.6	53.4	66.0	72.0
TOTAL (Mtoe)	5.3	33.6	46.5	57.5	65.3

Appendix II Table 5 **Natural Gas Demand by Sector: 1973-2010 Australia**
(billion cubic metres)

	1973	1984	1990	2000	2010
Low Demand/High Oil Price					
Industry	1.6	6.2	6.5	8.0	9.0
Residential/Commercial	1.0	2.2	2.5	3.2	3.7
Power Generation	0.9	3.1	3.0	2.3	2.0
Transport	-	-	-	-	-
Own Use and Losses	0.5	0.8	2.5	2.5	2.3
TOTAL (bcm)	4.0	12.3	14.5	16.0	17.0
TOTAL (Mtoe)	3.5	10.7	12.6	13.9	14.8
High Demand/Low Oil Price					
Industry	1.6	6.2	8.0	10.0	11.0
Residential/Commercial	1.0	2.2	3.0	4.0	4.0
Power Generation	0.9	3.1	4.0	3.5	3.0
Transport	-	-	-	-	-
Own Use and Losses	0.5	0.8	2.0	2.5	3.0
TOTAL (bcm)	4.0	12.3	17.0	20.0	21.0
TOTAL (Mtoe)	3.5	10.7	14.8	17.4	18.3

Appendix II Table 6 **Natural Gas Demand by Sector: 1973-2010 New Zealand**
(billion cubic metres)

	1973	1984	1990	2000	2010
Low Demand/High Oil Price					
Industry	0.04	1.28	1.4	1.7	1.7
Residential/Commercial	0.13	0.22	0.2	0.3	0.3
Power Generation	0.09	1.21	0.8	0.8	0.6
Transport	0	0.12	0.1	0.2	0.2
Own Use and Losses	0.07	0.14	0.2	0.2	0.2
Synthetic Fuels	0	0	1.3	1.3	1.3
TOTAL (bcm)	0.33	2.97	4.0	4.5	4.3
TOTAL (Mtoe)	0.29	2.59	3.5	3.9	3.7
High Demand/Low Oil Price					
Industry	0.04	1.28	1.4	1.9	2.1
Residential/Commercial	0.13	0.22	0.3	0.4	0.4
Power Generation	0.09	1.21	1.1	1.1	1.0
Transport	0	0.12	0.2	0.3	0.3
Own Use and Losses	0.07	0.14	0.2	0.2	0.2
Synthetic Fuels	0	0	1.3	1.3	1.3
TOTAL (bcm)	0.33	2.97	4.5	5.2	5.3
TOTAL (Mtoe)	0.29	2.59	3.9	4.5	4.6

Matching Supply and Demand: Netback Relationships

Average maximum burner-tip natural gas prices, competitive on a thermal equivalent basis with the relevant alternative fuels have been calculated for Western Europe. The energy price and sectoral demand forecasts outlined in Appendix II and described in Chapter IV above have provided the basis for these estimates which are shown in Appendix Table III-1. Distribution costs, estimated on an average basis and weighted for each sector of consumption have been:

  (i) added to ex-refinery oil product prices to yield ex-tax burner-tip prices for the relevant alternative fuels; and

  (ii) subtracted from the calculated average maximum competitive values to give a value for natural gas at the inlet of the transmission system (a "border value").

Where necessary, the distribution costs include an allowance for incremental storage investments.

The calculations, under the price, demand and cost assumptions of this study, suggest that new gas supplies would have to be capable of being developed in Europe at border prices of between \$2.10/MBTU and \$2.25/MBTU in 1990, rising to between \$3.75/MBTU and \$5.80/MBTU in 2000 and to between \$6.30/MBTU and \$7.60/MBTU by 2010. Different assumptions at any stage in the analysis (oil prices, volume and sectoral distribution of gas consumption, appropriate competing fuels, allocation of distribution costs and storage needs) would of course yield different values.

Appendix III Table 1
Natural Gas Burner-Tip and Netback Border Values in OECD Europe
(Constant 1984 $/MBTU)

	1984	1990	2000	2010
Weighted average "competitive fuel equivalent" value:	5.60	4.15-4.45	5.85-8.70	9.20-10.60
Weighted average[1] estimated distribution cost:	2.00	2.05-2.20	2.10-2.90	2.90-3.00
Weighted average Netback Border Value	3.60	2.10-2.25	3.75-5.80	6.30-7.60

1. Mid-point of range of weighted average distribution costs, including allowance for incremental storage investments where necessary.

Source: IEA Secretariat.

Among the potential suppliers of new gas to OECD European consumers are deep water North Sea fields and West African and Middle Eastern LNG projects. Appendix III Table 2 summarises the results of calculations of the raw economics of one "marginal" supply project from each of these areas. These economic calculations were done on a pre-tax, unleveraged basis. The gas prices assumed to calculate revenue streams are the netback border values indicated in Appendix III Table 1 above as consistent with the crude oil and product price assumptions in this study. Capital costs were assessed for the North Sea at $4.4 billion phased over seven years, including interest during construction. These are consistent with indicative figures for development of the Norwegian Troll field, to supply annual plateau volumes of 15 bcm of gas. For a small West African LNG project, supplying 4 bcm of gas per year, capital costs of just under $3 billion have been phased over seven years and used to include gathering, liquefaction and shipping costs. For the Middle East, for an 8 bcm per year LNG project, capital costs were assessed at $4.2 billion, phased over eight years. The figures are for gas-only economics. No credit was allowed for sale of any associated liquids or oil production; in the case of a field like the Troll field, which contains significant quantities (about 350 million barrels) of recoverable oil, the value of the oil would be an important part of any decision to develop the field. For the West African project, even though it is for a relatively small quantity of LNG, new regasification capacity was assumed to be required in Europe. If existing capacity were available, this could improve the economics.

The results in Appendix III Table 2 show positive real internal rates of return to all projects, against the real income streams based on both price scenarios. Sensitivities to a constant $25/bbl and $20.50/bbl oil price yield internal rates of return of about 13% and 10% respectively for the North Sea project (gas only economics). On this basis, it would appear possible to develop such projects profitably within the cost and competitive price parameters set. It may remain an open question as to whether the rates of return implied would exceed the thresholds required for commercial development given the risks involved.

These calculations are indicative, on the basis of assumptions in this study, of the raw economics of such potential new natural gas supplies.

Appendix III Table 2
Pre-Tax, Unleveraged Economics of Incremental Gas Supply Projects

	IRR	Net Present Value at 15% Real ($ million)	Net Present Value at 10% Real ($ million)
1. Deep-water North Sea Field			
Low Oil Price Case	17.0%	546	3 259
High Oil Price Case	20.7%	1 780	5 775
2. West African LNG			
Low Oil Price Case	9.4%	-534	- 64
High Oil Price Case	14.2%	- 78	799
3. Middle East LNG			
Low Oil Price Case	15.6%	72	1 169
High Oil Price Case	21.5%	854	2 713
4. Deep-water North Sea Field			
Constant $25/bbl crude oil price	13.1%	-403	1 098
Constant $20.50/bbl crude oil price	10.0%	-936	20

Source: IEA Secretariat.

The actual attractiveness to the producers, governments and financing institutions which could be involved will in fact depend on many other factors, as well as on the economic and risk-exposure implied, notably on proposed levels of government take and on the tax-paying position of the parties concerned.

APPENDIX IV

Glossary of Units and Conversion Factors

Unit	Abbreviation	Definition or Equivalent
1 barrel	bbl	= 42 United States gallons = 34.97 Imperial gallons = 158.99 litres
1 billion cubic feet [1]	bcf	= 0.0283 billion cubic metres [2]
1 billion cubic metres [2]	bcm	= 35.2913 billion cubic feet [1]
1 British Thermal Unit	BTU	= Heat required to raise the temperature of one pound of water through one degree Fahrenheit = 0.252 kilocalories = 1.055 kilojoules = 0.000293 kilowatt hours
1 Million British Thermal Units	MBTU	= 252×10^3 kilocalories = 1.055 Gigajoules = 293.071 kilowatt hours

1. 1 billion cubic feet at standard temperature and pressure conditions of 60ºF and 30"Hg.
2. 1 billion cubic metres at standard temperature and pressure conditions of 15ºC and 760mmHg.

Unit	Abbreviation	Definition or Equivalent
Calorific value	cv	= The calorific value of a fuel is the amount of heat in a given quantity of the fuel which can be released on combustion. Gross calorific value (gcv) is a measure of the total heat released on combustion; net calorific value (ncv) is the gross calorific value minus the latent heat of evaporation of water vapour released in the combustion process.
		The more hydrogen in the fuel (oil has more than coal, gas more than oil or coal), the more water vapour is formed and released on combustion, and the wider the difference between the gross and net calorific value.
Gross calorific value	gcv	See above.
1 kilocalorie	kcal	Heat required to raise one kilogram of water through one degree Celsius. = 3.96832 BTUs = 0.001163 kilowatt hours = 4.1868 kilojoules
1 million barrels per day	Mbd	= 49.5 million tonnes per year, for 34° API crude
1 million cubic feet per day	Mcfd	= 0.0283 Mcmd (at 15°C and 760mmHg)
1 million cubic metres per day	Mcmd	= 35.2913 Mcfd (at 60°F and 30"Hg)
1 million metric tons of oil equivalent	Mtoe	= 10^{13} kilocalories
Net calorific value	ncv	See under "calorific value" above.
Trillion cubic feet	Tcf	= 10^{12} cubic feet = 28.3 billion cubic metres
1 ton of oil equivalent	Toe	= 10^{7} kilocalories = 41.868 Gigajoules = 39.68 MBtus.

Conclusions on Gas: Meeting of Governing Board at Ministerial Level
8th May 1983

Ministers agreed that gas has an important role to play in reducing dependence on imported oil. They also agreed, however, on the importance of avoiding the development of situations in which imports of gas could weaken rather than strengthen the energy supply security and thus the overall economic stability of Member countries. They noted the potential risks associated with high levels of dependence on single supplier countries. Ministers stressed the importance of expeditious development of indigenous OECD energy resources. They noted that existing contracts are currently insufficient to cover expected gas demand by the mid-1990s, and agreed that in filling this gap steps should be taken to ensure that no one producer is in a position to exercise monopoly power over OECD and IEA countries. To obtain the advantages of increased use of gas on an acceptably secure basis, they agreed that:

- their countries would seek to avoid undue dependence on any one source of gas imports and to obtain future gas supplies from secure sources, with emphasis on indigenous OECD sources. Additional supplies from other sources would be obtained from as diverse sources as possible, taking into account supply structures, the share of gas in energy balances, and the geographical situation of individual countries. In assessing the full costs of gas supply sources, gas companies and, as appropriate, governments will consider security factors;

- their Governments would either encourage gas companies and other undertakings concerned to take or take themselves the necessary and appropriate cost-effective measures suited to each country's situation to strengthen their ability to deal with supply disruptions; these measures could include increased gas storage facilities, contingency demand restraint programmes, improved fuel-switching capabilities accompanied by adequate stocks of oil or other alternative fuels, a more flexible grid structure, greater flexibility of contracts, more surge capacity, measures to accelerate intra-OECD trade on short notice through standby contracts for supplies in a disruption, and interruptible contracts with consumers;

- action should be taken to develop at economic cost indigenous gas resources, particularly in North America and the North Sea, which show promise of alleviating overall or particular pressures on energy imports;

- concerned Member governments noting the potential for further development of North American gas resources and noting that part of the Norwegian Troll field may be declared commercial by 1984, would encourage their companies to begin negotiations on deliveries from these sources as soon as possible, with a view to making supplies available at prices competitive with other fuels in the mid-1990s;

- trade barriers and other barriers which could delay development of indigenous gas resources should be avoided or reduced;

- their governments would encourage the companies concerned to undertake feasibility studies, if appropriate in cooperation with Member governments, to determine the economic, engineering, technical and financial factors, relevant to possible imports from a variety of non-OECD sources;

- governments within one region where there is scope for effective cooperation should invite gas companies operating in their jurisdictions to address and negotiate on a commercial basis cooperative arrangements to meet a disruption of supplies to any one country or to the region as a whole;

- special attention should be given in the annual country review process in various international organisations to the future pattern of gas supplies, to the progress on the development and implementation of security measures, and to whether gas imports into the OECD from any single source constitute such a proportion of total supplies as to give rise to concern about the timely development of indigenous resources and the vulnerability of supplies, either for an individual Member country or collectively;

- in considering the degree of vulnerability, relevant factors include the share of imports in total gas consumption and in total primary energy requirements, the reliability of particular sources, the flexibility of other supplies, sectoral distribution, stocks and fuel-switching possibilities;

- an in-depth exchange of views about this question would take place within the normal review process whenever considered necessary. To allow a full assessment of its energy situation, the country concerned shall inform the other member states if it plans major changes in its energy policy or gas supply pattern which are significant in the context of development of indigenous OECD resources and vulnerability of gas supplies.

Ministers expressed the view that special attention should be given in relevant international organisations to the gas import situation of individual countries and regions. IEA Ministers instructed the Governing Board to keep this issue under continuing review.

ANNEX II

Conclusions of the Governing Board of the IEA Meeting at Ministerial Level, 9th July 1985 Concerning Natural Gas

Ministers agreed that the following actions are required for implementation of the May 1983 Conclusions:

- Avoidance of undue dependence on any one source of gas imports and obtaining future gas supplies from secure sources, with emphasis on indigenous OECD sources. Additional supplies from other sources should be obtained from as diverse sources as possible, taking into account supply structures, the share of gas in energy balances, and the geographical situation of individual countries. In assessing the full costs of gas supply sources, gas companies and, as appropriate, governments will consider security factors.

- Development of indigenous gas resources in particular in North America and the North Sea, including the Norwegian Troll field, with a view to making supplies available at prices competitive with other fuels in the mid-1990s.

- Necessary and appropriate cost-effective measures suited to each country's situation to strengthen their ability to deal with supply disruptions.

- Measures to avoid an increased reliance on oil if gas supplies should prove inadequate to meet demand.

Ministers noted that the Secretariat is updating the study on Natural Gas Prospects to 2000 published in 1982. They requested the Governing Board at official level to review the results of this work and to draw any necessary policy conclusions from it.

OECD SALES AGENTS
DÉPOSITAIRES DES PUBLICATIONS DE L'OCDE

OECD PUBLICATIONS, 2, rue André-Pascal, 75775 PARIS CEDEX 16 - No. 43631 1986
PRINTED IN FRANCE
(61 86 06 1) ISBN 92-64-12822-0

INTERNATIONAL ENERGY AGENCY

INTERNATIONAL ENERGY AGENCY

2, RUE ANDRÉ-PASCAL 75775 PARIS CEDEX 16, FRANCE

The International Energy Agency (IEA) is an autonomous body which was established in November 1974 within the framework of the Organisation for Economic Co-operation and Development (OECD) to implement an International Energy Program.

It carries out a comprehensive programme of energy co-operation among twenty-one* of the OECD's twenty-four Member countries. The basic aims of IEA are:

  i) co-operation among IEA Participating Countries to reduce excessive dependence on oil through energy conservation, development of alternative energy sources and energy research and development;
 ii) an information system on the international oil market as well as consultation with oil companies;
iii) co-operation with oil producing and other oil consuming countries with a view to developing a stable international energy trade as well as the rational management and use of world energy resources in the interest of all countries;
 iv) a plan to prepare Participating Countries against the risk of a major disruption of oil supplies and to share available oil in the event of an emergency.

IEA Member countries: Australia, Austria, Belgium, Canada, Denmark, Germany, Greece, Ireland, Italy, Japan, Luxembourg, Netherlands, New Zealand, Norway, Portugal, Spain, Sweden, Switzerland, Turkey, United Kingdom, United States.

Pursuant to article 1 of the Convention signed in Paris on 14th December, 1960, and which came into force on 30th September, 1961, the Organisation for Economic Co-operation and Development (OECD) shall promote policies designed:

 – to achieve the highest sustainable economic growth and employment and a rising standard of living in Member countries, while maintaining financial stability, and thus to contribute to the development of the world economy;
 – to contribute to sound economic expansion in Member as well as non-member countries in the process of economic development; and
 – to contribute to the expansion of world trade on a multilateral, non-discriminatory basis in accordance with international obligations.

The Signatories of the Convention on the OECD are Austria, Belgium, Canada, Denmark, France, the Federal Republic of Germany, Greece, Iceland, Ireland, Italy, Luxembourg, the Netherlands, Norway, Portugal, Spain, Sweden, Switzerland, Turkey, the United Kingdom and the United States. The following countries acceded subsequently to this Convention (the dates are those on which the instruments of accession were deposited): Japan (28th April, 1964), Finland (28th January, 1969), Australia (7th June, 1971) and New Zealand (29th May, 1973).

The Socialist Federal Republic of Yugoslavia takes part in certain work of the OECD (agreement of 28th October, 1961).